Lab Manual to Accompany the Electronics Trainer Circuit Board (eTCB) System

by Carlo Sapijaszko

THOMSON
DELMAR LEARNING

Lab Manual to Accompany the Electronics Trainer Circuit Board (eTCB) System

Carlo Sapijaszko

Vice President, Technology and Trades SBU:
Alar Elken

Editorial Director:
Sandy Clark

Senior Acquisitions Editor:
Stephen Helba

Senior Channel Manager:
Dennis Williams

Senior Development Editor:
Michelle Ruelos Cannistraci

Marketing Director:
Dave Garza

Marketing Coordinator:
Stacey Wiktorek

Production Director:
Mary Ellen Black

Production Manager:
Andrew Crouth

Senior Project Editor:
Christopher Chien

Art/Design Coordinator:
Cheri Plasse

Technology Project Manager:
Kevin Smith

Senior Editorial Assistant:
Dawn Daugherty

Printed in the United States of America
1 2 3 4 5 XX 06 05

For more information contact
Thomson Delmar Learning
Executive Woods
5 Maxwell Drive, PO Box 8007,
Clifton Park, NY 12065-8007
Or find us on the World Wide Web at
www.delmarlearning.com

ISBN: 1-4018-3980-0

NOTICE TO THE READER

Publisher does not warrant or guarantee any of the products described herein or perform any independent analysis in connection with any of the product information contained herein. Publisher does not assume, and expressly disclaims, any obligation to obtain and include information other than that provided to it by the manufacturer.

The reader is expressly warned to consider and adopt all safety precautions that might be indicated by the activities herein and to avoid all potential hazards. By following the instructions contained herein, the reader willingly assumes all risks in connection with such instructions.

The publisher makes no representation or warranties of any kind, including but not limited to, the warranties of fitness for particular purpose or merchantability, nor are any such representations implied with respect to the material set forth herein, and the publisher takes no responsibility with respect to such material. The publisher shall not be liable for any special, consequential, or exemplary damages resulting, in whole or part, from the readers' use of, or reliance upon, this material.

Contents

Introduction

The **Electronics Trainer Circuit Board (eTCB) System: DC/AC Module** consists of the following components:

- eTCB custom board that works with National Instruments' NI ELVIS (Electronics Laboratory Virtual Instrumentation Suite) (ISBN: 1401839797)
- eTCB software (ISBN: 1401839797)
- eTCB lab manual with over 30 DC/AC lab experiments (ISBN: 1401839800)

Combining virtual and physical instrumentation, the eTCB System provides a powerful, cost-savings solution for students to perform real-world DC/AC experiments. Just plug eTCB into NI ELVIS and a personal computer, and you've got a complete electronics laboratory at your fingertips.

Innovative and pedagogically rich, the system was developed to achieve the following goals:

- To ensure that all of the necessary circuit configurations, using real and practical industry circuit fulfill the needs of any DC/AC course.
- To ensure that students who perform DC/AC experiments focus on achieving the objectives of a given laboratory experiment.
- To ensure that students experiment with real components mounted on an industry-standard PCB.
- To ensure that students are themselves physically placing test leads to perform measurements, thus developing industry-relevant experience.
- To ensure that students are measuring real electrical quantities, and do not rely exclusively on simulation for their laboratory work

Features:

1. The eTCB system achieves the above goals in part by providing a controlled environment where enough flexibility exists for students to make their mistakes and learn from them, and enough control exists to keep students focused on the essence of each of the experiments. With the eTCB System, a traditional computer lab, equipped with PCs and desks, can be transformed into an electronics laboratory simply by deploying the eTCB system.

2. The lab manual and eTCB form a complete package that does not require students to add extra components. The eTCB was designed in such a way as to replace the functionality of a traditional breadboard, with the concept of a configurable circuit.

3. Well-chosen configurable circuit blocks have been embedded in the eTCB, and the connections between the various circuit blocks are made with relay contacts driven by eTCB software. With NI ELVIS seamlessly integrated into this solution, students have access to an array of instruments, including a DMM, an oscilloscope, a function generator, a variable power supply, a dynamic signal analyzer, and more. This solution ensures that students can perform laboratory experiments with real components, manually measure real electrical quantities, and work with industry-standard designs.

Overview of the eTCB System

System Requirements

To set up and use the eTCB System, the following are required:

- The eTCB (elecronics trainer circuit board)
- The eTCB Software
- The eTCB Lab Manual

- The National Instrument's ELVIS workstation, data acquisition board, and LabView 7.0 or later
    - To order ELVIS, visit ni.com/academic for information on products, curriculum resources, and special academic discounts.
    - Refer to NI ELVIS user's guide for set-up instructions.
- A computer
- Windows NT/2000/XP
- Recommended 1280 X 1024 resolution with true
    - Color
- 256 MB RAM
- CD-ROM drive
- One PCI or PCMCIA card slot

This system is self-contained; no other electrical components or hardware is required.

- The function of the computer in the eTCB system is to provide a graphical user interface for all of the instruments programmed into the eTCB software application.
- The purpose of the eTCB board is to provide a platform on which students can perform laboratory experiments. The eTCB board contains configurable circuits, which are appropriate for a given curriculum. Its purpose is also to communicate circuit configuration with the PC.
- The eTCB lab manual offers an exhaustive array of laboratory experiments for a given curriculum.
- Refer to Figure 1 and 2 on setting up the eTCB System.

Figures 1 and 2 provide details of the eTCB, as shown.

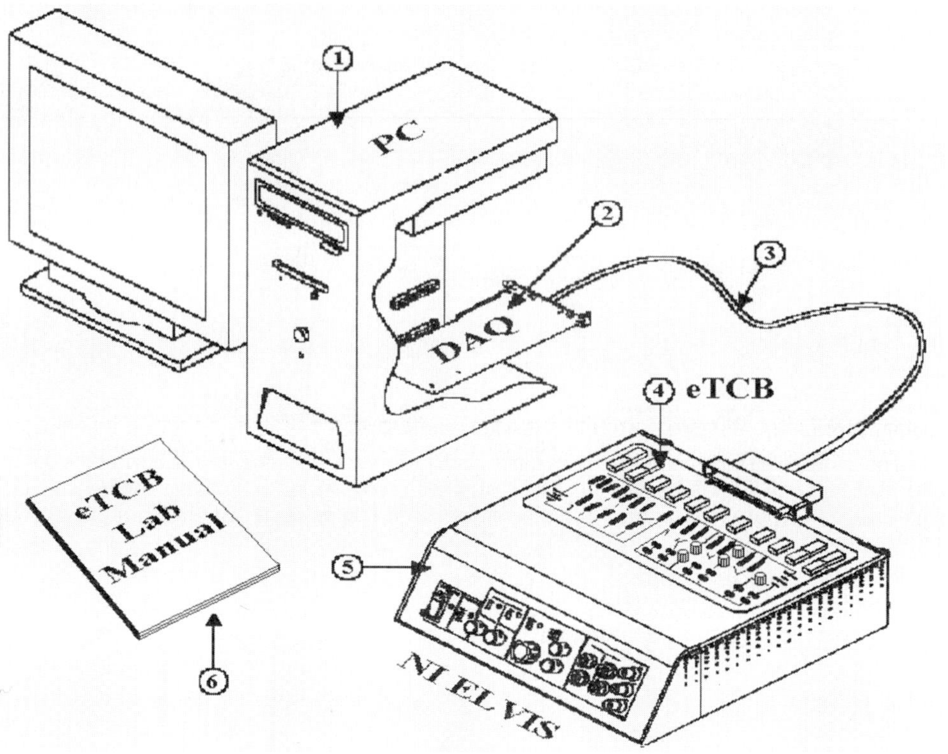

Figure 1
1. PC, 2.DAQ Device, 3. DAQ Cable, 4. eTCB Board, 5. NI ELVIS Workstation, 6. eTCB Lab Manual

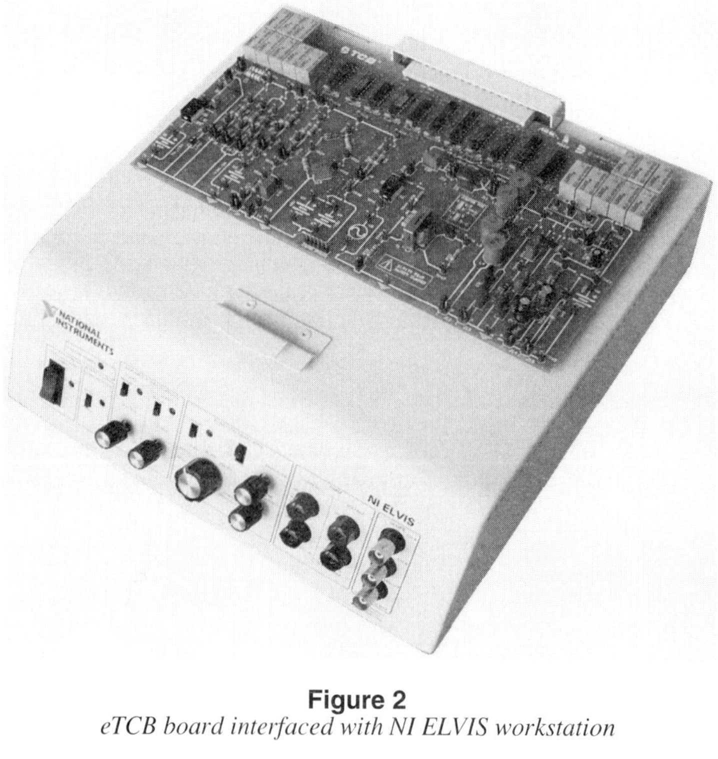

Figure 2
eTCB board interfaced with NI ELVIS workstation

Performing Self-Diagnostic

Once the eTCB System is set up, it is important to utilize the self-diagnostic feature. The purpose of the self-diagnostic feature built into the eTCB system is to assure the student that the eTCB system is performing within specified tolerances. This information is critically important for students to keep their focus on performing the experiment and to prevent them from unnecessarily questioning the platform on which they perform the experiment. Please refer to the list of diagnostic codes at the end of the lab manual.

In order for the system to start performing its self-diagnostic, all the conditions for turning the No/Yes switches need to be satisfied. (Refer to Figure 3 for location of the switches.) Only then the No/Yes switch can be switched to a Yes position. Once all No/Yes switches are in the Yes position the system will automatically start performing its self-diagnostic.

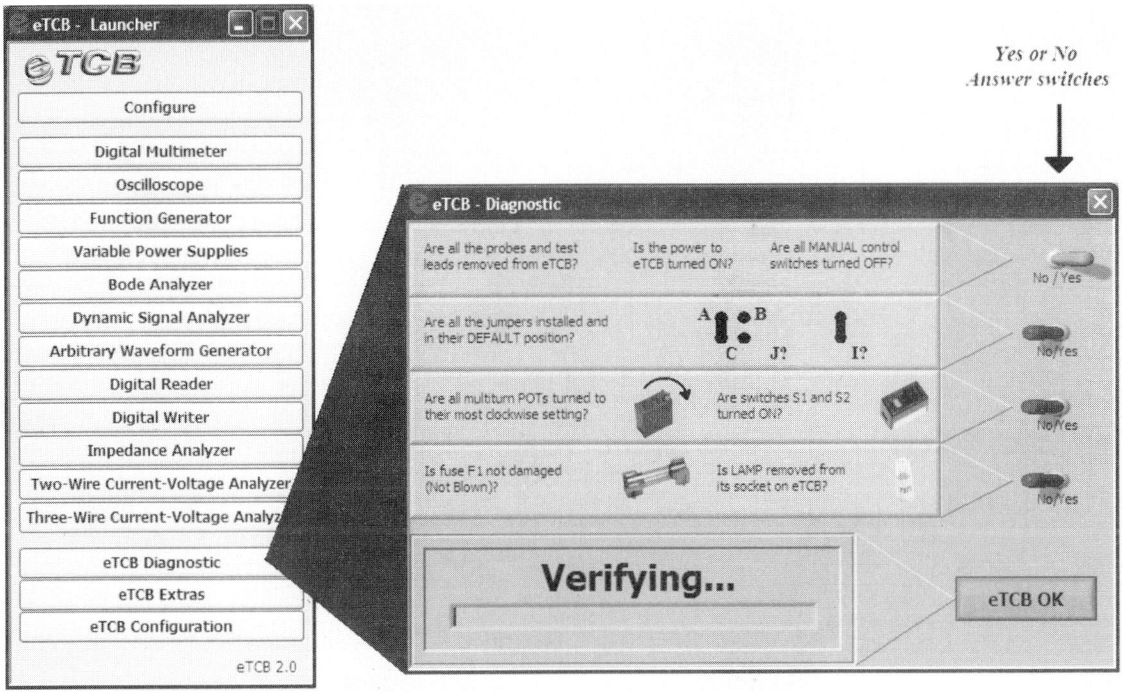

Figure 3

Performing a self-diagnostic on eTCB system.

Prior to performing the self-diagnostic, it is necessary that all J jumpers are installed in their default position between terminal A and C. In some instances this may position the jumper horizontally. Some I jumpers are placed on to a two pin terminals; those jumpers must all be installed.

Three Manual selections switches are located on the front panel of the NI ELVIS workstation: two for variable power supplies and one for function generator. All of those switches are to be turned to OFF (placed down).

Configuring Circuits

The eTCB board has been designed with the concept of configurable circuits as a means of building experiment circuits. This approach eliminates the need for breadboards as a means of constructing electrical circuits. This alleviates the difficulties of debugging student-constructed circuits in an environment where direct supervision by an instructor is not available. Circuits to be examined during a given laboratory experiment are configured into the eTCB board with the use of a personal computer. Once the configuration is loaded into the eTCB board buffer, appropriate relays are turned to ON and their contacts make connections between various circuit blocks, thus configuring the circuits.

A schematic diagram of the DC module is shown in Figure 4. It illustrates the available DC circuit blocks from which specific circuits can be configured.

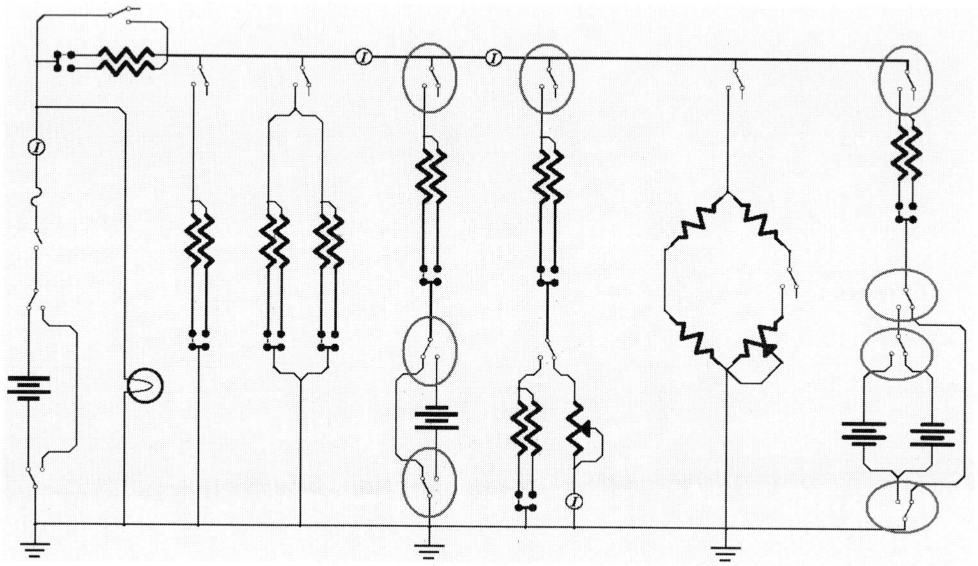

Figure 4
Circuit building blocks for the DC module

By turning ON the circled relays shown in Figure 4, the circuit shown in Figure 5 is configured. In this manner, circuits of varying complexity can be configured by selecting appropriate relays to turn ON or OFF.

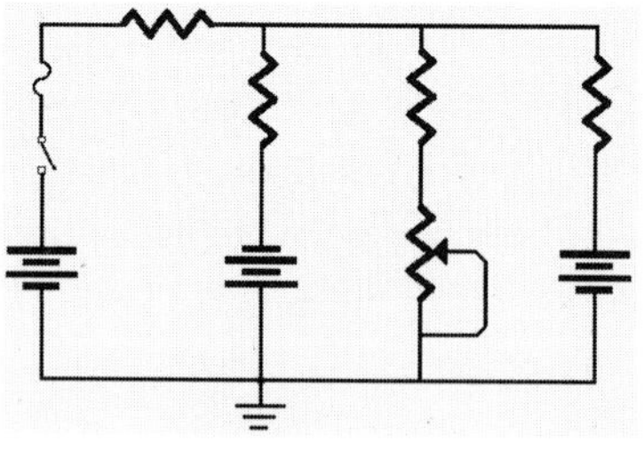

Figure 5
Configured DC circuits

As part of every experiment procedure, students are asked to configure an appropriate experiment circuit. Although Figure 4 shows DC voltage sources and resistive loads only, other modules address the need for AC voltage sources, and passive components such as capacitors and inductors.

Figure 6 shows configurable circuits addressing the need for an AC voltage source and other passive components.

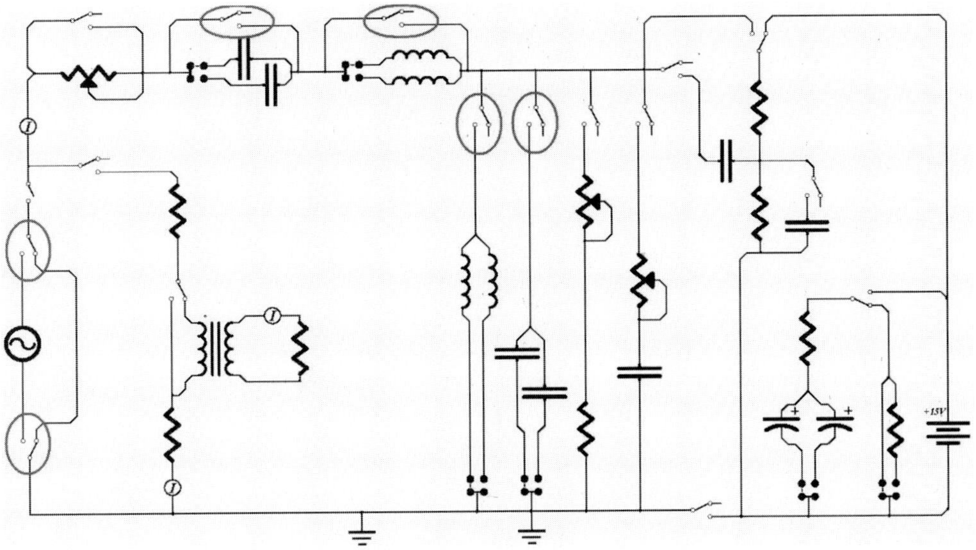

Figure 6
Circuit building blocks for the AC module

By turning ON the circled relays shown in Figure 6, the circuit shown in Figure 7 is configured. In this manner, circuits of varying complexity can be configured by selecting appropriate relays to turn ON or OFF.

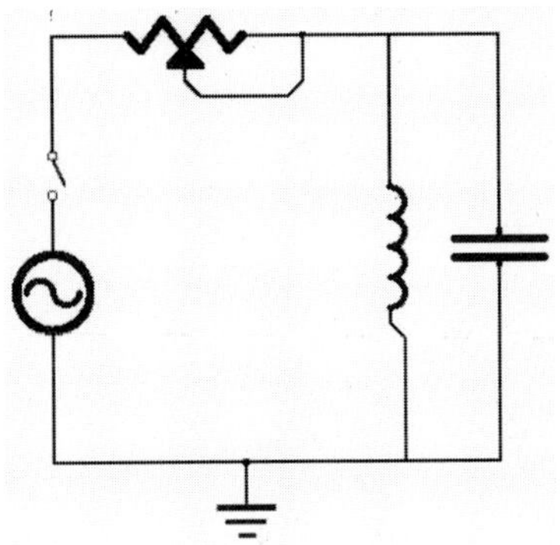

Figure 7
Configured AC circuits

In order to add further flexibility to the multitude of circuits that can be configured, most of the components come as pairs. Components in pairs differ from each other by their ratings. As an example, a resistor pair is made up of 470 Ω resistor and a 820 Ω resistor. Components are selected by using a physical jumper, which engages one of the two components in a pair. Another possibility is to place two jumpers, engaging both components at the same time.

Figure 8 illustrates the concept of jumper-selectable components.

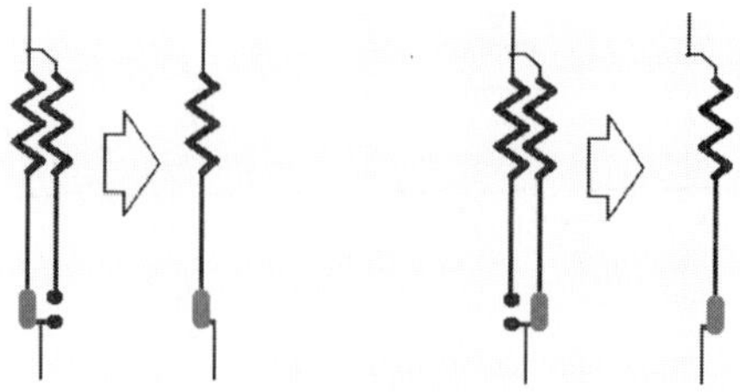

Figure 8
Jumper-selectable components

All circuit configurations are performed with the eTCB Configuration instrument available from the eTCB Launch panel. Clicking on the icon representing a relay contact turns ON a physical relay located on the eTCB board. This action also highlights the portion of the circuit that has just been connected to the circuit.

Figure 9 illustrates how a five-resistor and three-DC voltage supplies circuit is configured using the eTCB Configuration instrument. With the same process, circuits of varying complexity can be configured by selecting appropriate relays to turn ON or OFF.

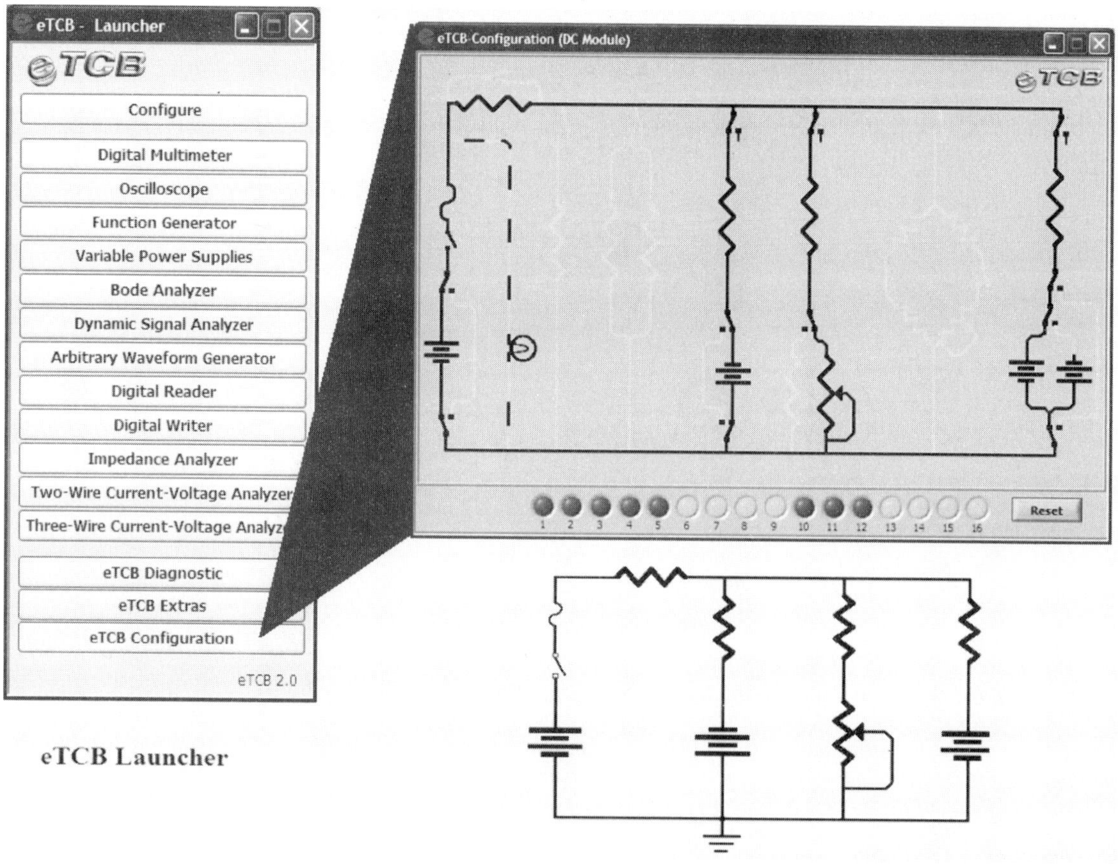

Figure 9
Configuring a DC circuit using eTCB Configuration instrument

Placing Test Probes on eTCB

The eTCB board is designed with many test points strategically placed to allow for various measurements to be obtained. Jumpers, which are used for selecting components from their pairs, are also used to break the circuits and allow for current measurements. Refer to Figure 10.

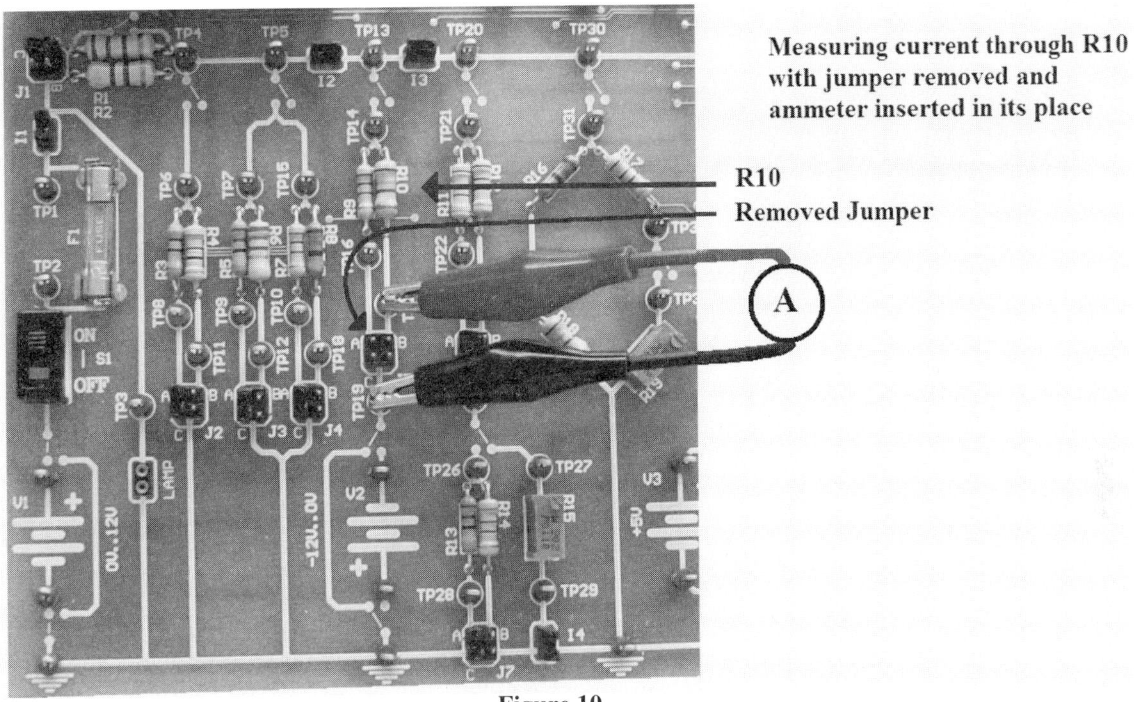

Figure 10
The use of component-selecting jumper ports in current measurements

The DC/AC board features the strong geographical similarities of the schematic diagram layout and the actual layout of the components on the printed circuit board (PCB). (See Figure 11)

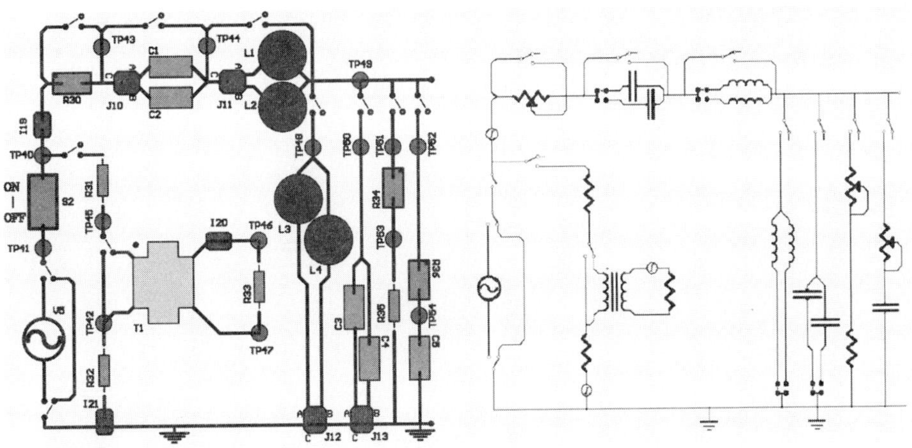

Figure 11
Strong similarities of the geographical layout of the PCB and the schematic diagram.

Working with Real Components and Real Measurements

The eTCB board is designed to give students their first exposure to working with real components mounted on a printed circuit board (PCB). Students are expected to obtain experiment measurements by connecting test leads to existing test points found on the eTCB board. The design uses the following components shown in Figure 12:

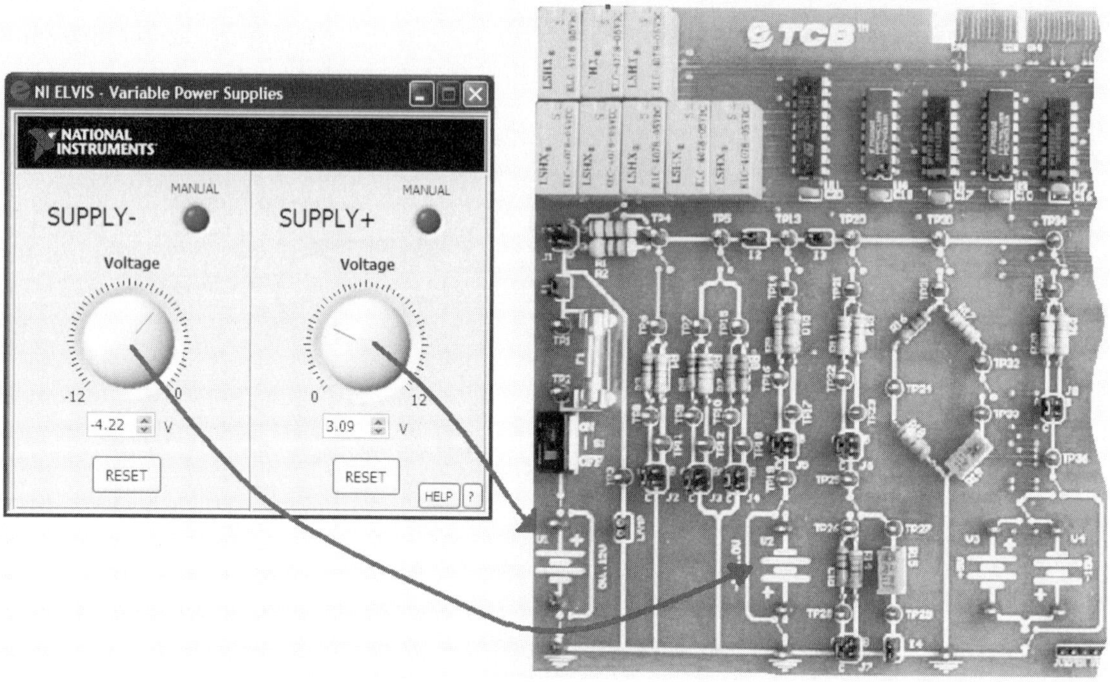

Figure 12

Mapping of the variable power supplies to eTCB board layout

With the use of available instruments from the eTCB launcher, students can perform a variety of measurements. The DMM test leads and oscilloscope probes can be connected to the terminals on the front panel of the NI ELVIS workstation. These leads and probes can then be connected to the various test points on the eTCB board in order to obtain measurements. None of the electrical quantities are simulated; all components and electrical signals are real. Measurements of these electrical quantities can be performed either with stand-alone test instruments or (preferably) with the eTCB instruments. In either case, measurements are performed on real electrical quantities.

Figure 13 illustrates how DMM test leads connected to the front panel of an NI ELVIS workstation that has been set to operate as an eTCB DMM, serve the same function as a stand-alone physical DMM.

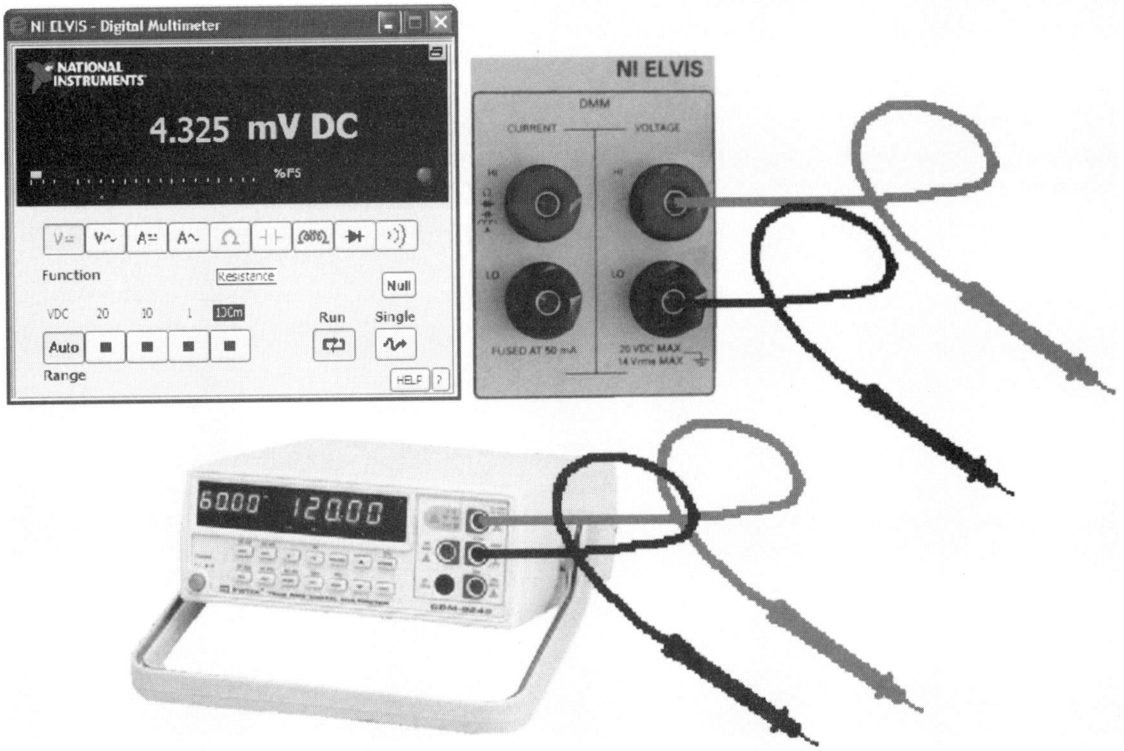

Figure 13
Similarities between a virtual eTCB instrument and a physical stand-alone instrument

Lab Manual

The Lab Manual includes 32 DC/AC experiments using the eTCB system and fulfills the requirements of any DC/AC course by taking the student from the basics of engineering notation all the way to passive filter analysis and design.

Each lab features:

- **Problem Statement** – presents a practical application.
- **Concept** – reviews the theoretical background of the experiment's content.
- **Learning Objectives** – outlines a list of practical skills to be learned by the students.
- **Pre-Lab Procedure** – requires students to perform theoretical work which is to be confirmed later in the experiment.
- **Lab Procedure** – outlines a specific set of instructions for students to follow. The steps in this section are designed to be easily verifiable. This approach allows the computer to render immediate feedback to students performing the experiment.
- **Questions** – enables students to further their understanding by solving practical problems and further analyzing experiment data.
- **Troubleshooting** – enables students to practice their critical thinking skill by troubleshooting faulty circuits.

Online Companion

Visit www.delmarlearning.com/etcb for updates and additional information.

Acknowledgments

The Author and Thomson Delmar Learning would like to thank the following people for their insight during the development of the eTCB System:

Eric Dean and Ravi Marawar from National Instruments, Greg Clayton from Thomson Delmar Learning, Seddik Benhamida, Michael Keresztesy, Scott Porter, and Isabel Perez from DeVry University, Doss Sowri from Orange County School, Larry Schultz from Missaukee Career Technical Center, Takis Kasparis from University of Central Florida, Philip Schmidt from University of Texas, Vladimir Genis and Shodhan Shetty from Drexel University.

1

Engineering/Scientific Notation and Symbols

Problem Statement

A digital multimeter is a measuring instrument used to measure various electrical quantities. Consider a display of such an instrument, as shown in Figure 1-1

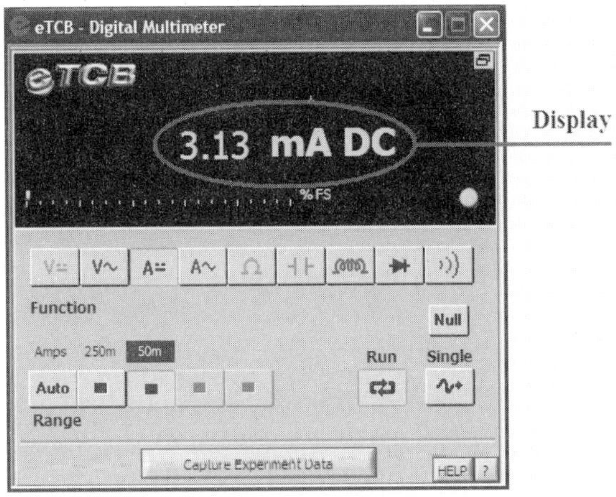

Figure 1-1

In this case, the value of measured electrical quantity is 0.00313 A. However, with only three display digits available, it would be impossible to display the measured value. The solution to this challenge is to convert the decimal number into engineering notation. Display of very large or very small values becomes much more practical with the use of engineering notation.

Concept

A large part of the study of electronics involves working with very small or very large numbers; in their generic form, these numbers become very complicated to work with and to identify. Using scientific notation and engineering notation makes solving problems with extremely large or small numbers manageable. Engineering notation is based on scientific notation with the addition of metric prefixes rather than exponential powers of 10. Most calculators in use today recognize engineering notation.

Besides the expression of mathematical concepts, it is necessary to describe various physical aspects of electrical or electronic circuitry in pictorial form. Standard electronic component symbols have been developed for this purpose. They allow the engineers to express circuit characteristics in an understandable, efficient manner.

Learning Objectives

Upon completion of this lab the student will be able to:

1. Convert decimal numbers to scientific notation.
2. Convert numbers from scientific notation into decimal form.
3. Express large and small numbers in engineering notation prefixes.
4. Identify various electronic symbols.

Pre-Lab Procedure

1. Complete Table 1-1 by drawing a schematic symbol for the given electrical component.

Table 1-1

	Schematic Symbol
DC Voltage Source	
AC Voltage Source	
Resistor	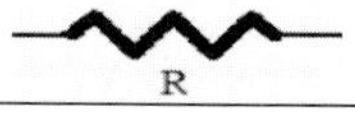
Capacitor	
Inductor	
Potentiometer (POT)	
Lamp	
Fuse	
SPST switch	
SPDT switch	
DPDT switch	
Signal Ground	
Chassis Ground	

2. Complete Table 1-2 by writing units of measure (complete words) and abbreviation for units of measure for the given electrical quantities.

Table 1-2

	Unit of Measure	***Abbreviated Unit of Measure.***
Voltage	*Volt*	*V*
Current		
Power		
Energy		
Resistance		
Impedance		
Conductance		
Reactance		
Capacitance		
Inductance		
Frequency		
Time		
Charge		

3. Complete Table 1-3 by converting decimal numbers into scientific notation with two decimal places. (Round to the nearest number if necessary.)

Table 1-3

Decimal Number	***Scientific Notation***
4,563	4.56×10^{3}
45,000	
98,467,000,000	
89	
5	
0.450	
0.000,0435	
0.000,000,000,034,53	

4. Complete Table 1-4 by converting decimal numbers into engineering notation with two decimal places. (Round to the nearest number if necessary.)

Table 1-4

Decimal Number	*Engineering Notation*
45,630	*45.63 x10³*
67,397,000	
702,400,000,000	
89	
5	
0.450	
0.000,0434,6	
0.000,000,000,034 5	

5. Complete Table 1-5 by converting decimal numbers into engineering notation with units of measure. (Round to the nearest number if necessary.)

Table 1-5

Decimal Number	*Engineering Notation*
76,830 V	*76.83k V*
47,590,000 _	
8,702,000,000 W	
49 J	
2 A	
0.000,038 H	
0.000,000,074,54 F	

6. Complete Table 1-6 by converting engineering notation with units of measure into decimal numbers.

Table 1-6

Engineering Notation	*Decimal Number*
45k V	*45,000 V*
5.2M _	
898G W	
86 J	
220p F	
47m H	
680_ A	

Lab Procedure

Step#1 Using the eTCB Extra VI (Notation), verify results from Table 1-3.

Step#2 Using the eTCB Extra VI (Notation), verify results from Table 1-4.

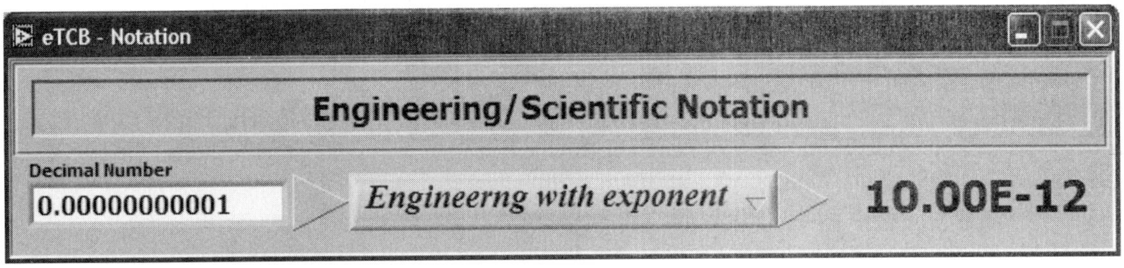

Step#3 Using the eTCB Extra VI (Notation), verify results from Table 1-5.

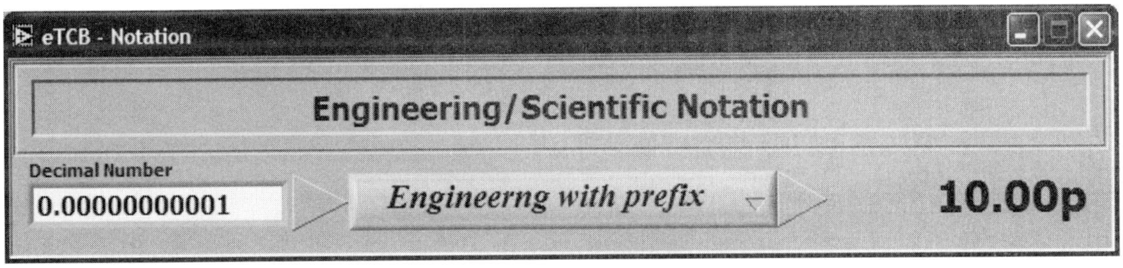

This concludes the procedure for this experiment.

Questions

1. What is the advantage of using scientific notation over decimal numbers?

2. What is the advantage of using engineering notation over scientific notation?

Bonus:

How many bytes are there in 1M byte memory chip? Be careful, the answer may not be obvious.

2

Open and Closed Circuits

Problem Statement

The understanding of the basic building blocks of electrical circuits is one of the most fundamental skills that you need to master. Consider a household electrical system; it contains elements to control current flow, elements to perform useful work, elements to conduct current flow, and an energy source. Understanding the purpose of each of those elements will enable you to construct, analyze, and troubleshoot electrical systems.

Concept

Energy can neither be created nor destroyed. It can only be converted from one form to another. When we use electrical energy, we are doing work by converting it to another form of energy. We must have a circuit in order for electrical work to be done. A circuit consists of four parts: the energy source; the load or work to do; a path from the energy source to the load; and a return path from the load to the energy source. (Refer to Figure 2-1.)

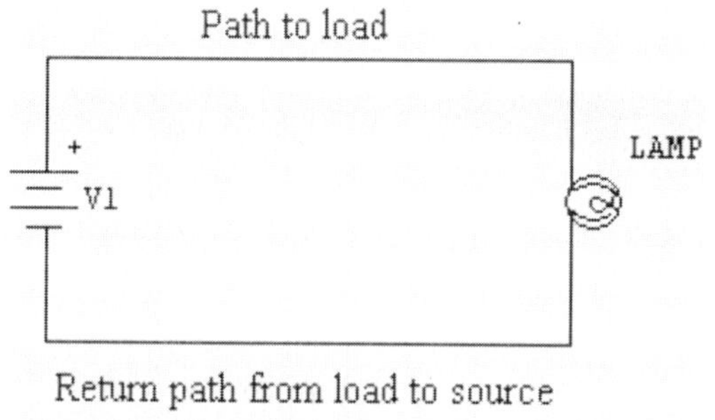

Figure 2-1

When work is done in a circuit, switches are devices designed to control. When there is a complete circuit, work is done. When the circuit is not complete, no work is done. Switches control when work is done by opening or closing a circuit. Circuits may be controlled from more than one position, depending on the type of switch used.

Circuit #1 in Figure 2-2 shows a scenario where the circuit is open and no work is done. The switch S1 in Figure 2-2 is known as single-pole-single-through (SPST). It can open or close connection between its two terminals.

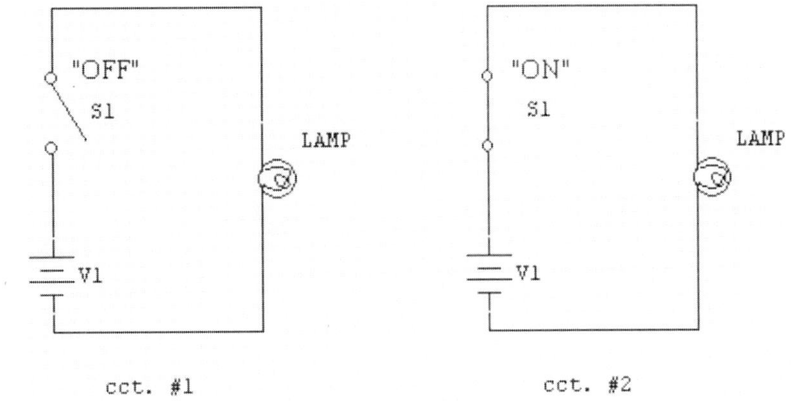

Figure 2-2

Circuit #1 in Figure 2-3 shows this scenario: the circuit is closed and work is done with lamp 1, and the circuit is open and no work is done with lamp 2. This is because switch S1 closes the path for current to flow from source to lamp 1 and opens (breaks) the path for current to flow to lamp 2. Circuit #2 shows a scenario where the circuit is open (broken) and work is not done with lamp 1, and the circuit is closed and work is done with lamp 2. This is because switch S1 opens (breaks) the path for current to flow from source to lamp 1 and closes the path for current to flow to lamp 2. The switch S1 in Figure 2-3 is known as single-pole-double-through (SPDT).

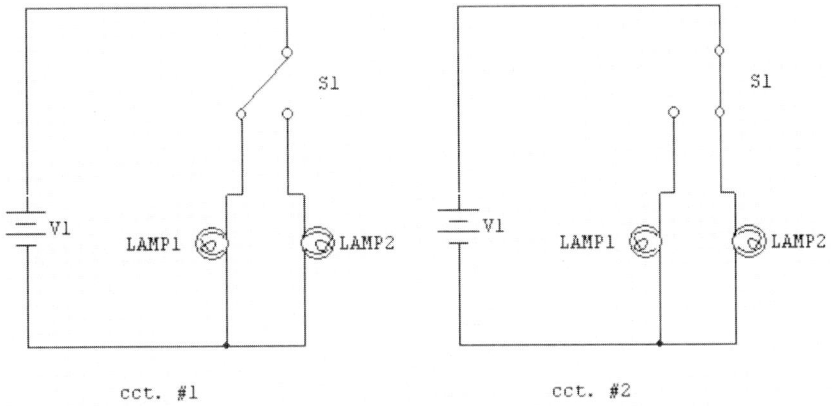

Figure 2-3

Switches such as SPST and SPDT can be actuated (controlled) mechanically or electro-mechanically. An example of a mechanically actuated SPST switch is a traditional light switch in a house. You mechanically turn the switch "ON" or "OFF". An example of an electro-mechanically actuated switch is a relay found in an engine starting circuitry on a car. In this situation a relatively small electrical signal controlled by the ignition switch is electrically turning the relay "ON". That relay being electrically actuated is then mechanically closing the circuit for a large current to flow from the source (car battery) to the engine starter motor.

Figure 2-4 shows a simplified schematic diagram of an engine starting circuitry.

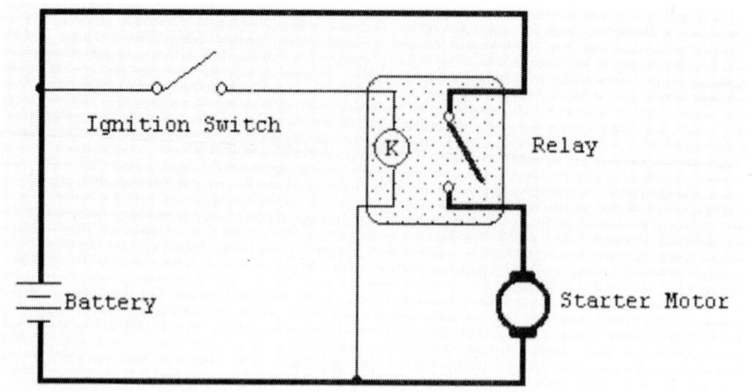

Figure 2-4

Learning Objectives

Upon completion of this lab, the student will be able to:

1. Configure a simple circuit using a schematic diagram.
2. Determine if a circuit has a complete path for current to flow.
3. Control a lamp with a switch.
4. Control a lamp with a relay.

Pre-Lab Procedure

1. In both circuits in Figure 2-2 identify and highlight the energy source, the load, a path from source to load, and return path from load to source.

2. Use Figure 2-6 to determine which switch and relay needs to be "ON" and which needs to be "OFF" in order to close the circuit and have it complete with all four circuit parts (source, load, path to load, return path).

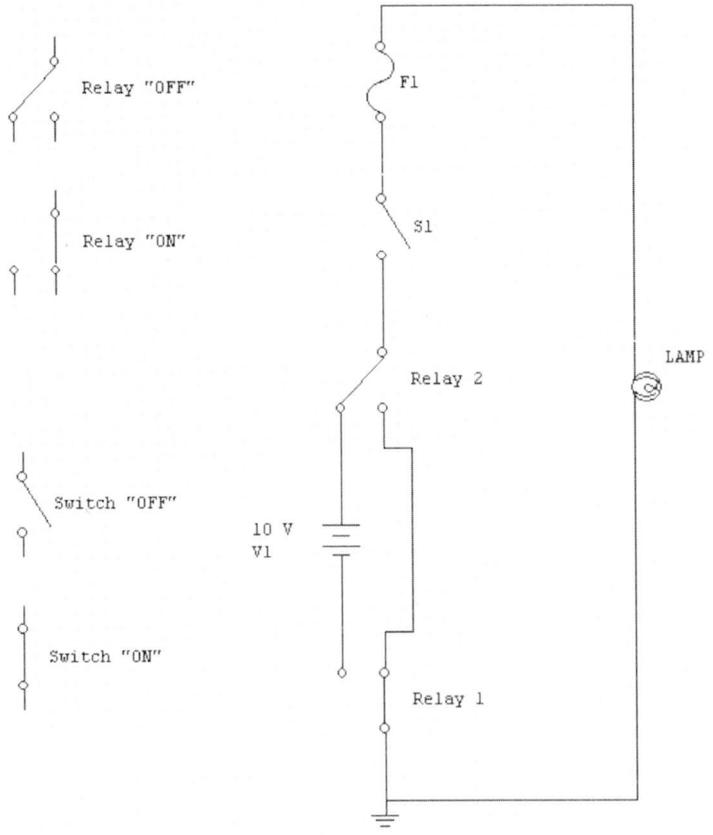

Figure 2-6

Lab Procedure

Before proceeding further, make sure that your eTCB is working correctly. Using eTCB Diagnostic VI, perform a self-diagnostic of eTCB circuitry.

Turn the switch S1 and S2 OFF in preparation for the lab.

Setup

Step #1 Insert the lamp in its socket on the eTCB board. Inserting the lamp will complete the dashed line connection. Using the eTCB Configuration VI, (DC Module), configure the circuit in Figure 2-7. This circuit is not yet complete, as the path for the current is not yet made.

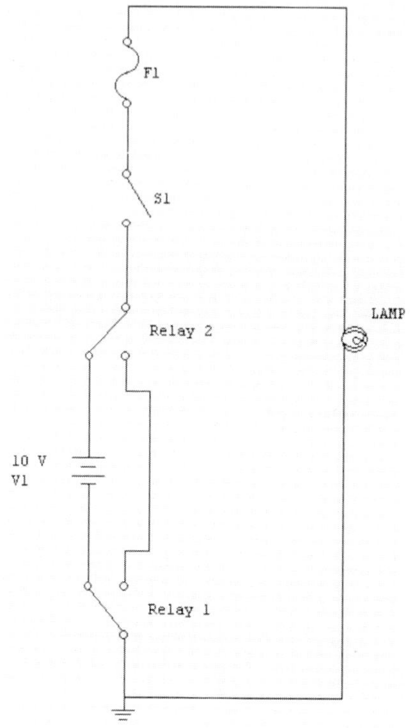

Figure 2-7

Step #2 Using the Variable Power Supplies VI, set voltage source V1 to 10V. Keep the Variable Power Supplies VI ***open***; if you close it, all the variable voltage supplies will reset to 0V.

Discovery

Step #3 Physically turn the switch S1 on the eTCB board "ON". This will complete your circuit. Turn the switch S1 "OFF" and "ON" to complete the following table:

Switch S1	***Lamp***
"ON"	Lit (yes / no)
"OFF"	Lit (yes / no)

Step #4 With the switch S1 in the "ON" position, use the eTCB Configuration VI, to turn relay 1 "ON". Turn relay 1 "OFF" and "ON" to complete the following table:

Relay 1	***Lamp***
"ON"	Lit (yes / no)
"OFF"	Lit (yes / no)

Step #5 With the switch S1 in the "ON" position, use the eTCB Configuration VI, to turn relay 1 "OFF" and to turn relay 2 "ON". Turn relay 2 "OFF" and "ON" to complete the following table:

Relay 2	***Lamp***
"ON"	Lit (yes / no)
"OFF"	Lit (yes / no)

This concludes the procedure for this experiment.

Questions

1 What four circuit parts must be present in order to have a complete circuit?

2 Identify and highlight all four parts of a complete circuit in circuit #1 and circuit #2 of Figure 2-8.

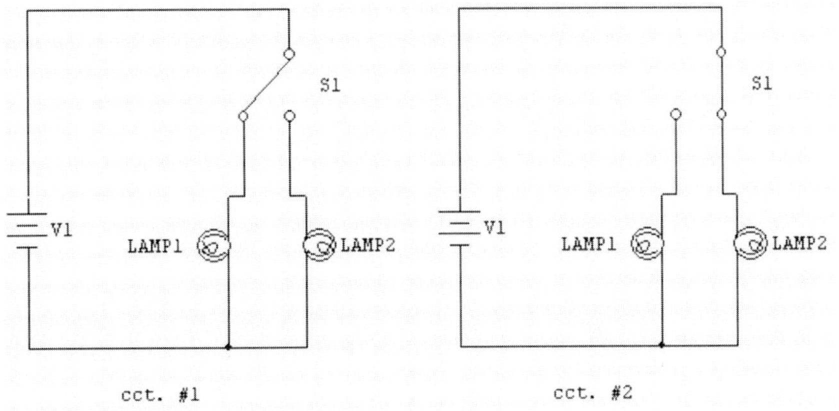

Figure 2-8

Bonus:

3 Complete the connections in circuit in Figure 2-9 to build a Two-Way-Switch system. A Two-Way-Switch system is guaranteed to turn the lamp "ON" or "OFF" by controlling either of the two switches.

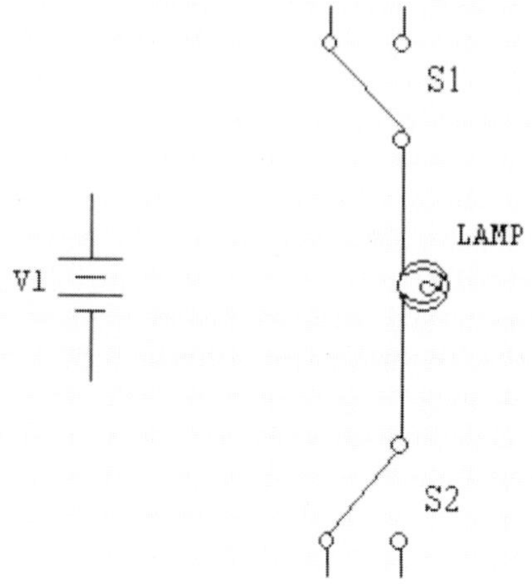

Figure 2-9

3

Current

Problem Statement

Electrical current refers to a movement of electrical charge. By understanding how to control the current flow one can design, analyze, and troubleshoot electrical circuits. Consider a typical household space heater: by understanding what current is and how to control it, you can analyze and troubleshoot simple appliances like a household heater.

Concept

Current (I) is the rate of flow of electric charge. If one coulomb of charge passes a given point in one second, there is one amp of current flow. Current is measured in amperes, or amps. The symbol for the amp is "A". Current is the quantity being measured, while amps are the unit of measurement. Current may be measured with an ammeter. An ammeter is inserted in series with the component through which current is to be measured, as shown in Figure 3-1. Ideally an ammeter represents a short, or path of zero resistance. This keeps the meter from affecting the operation of the circuit.

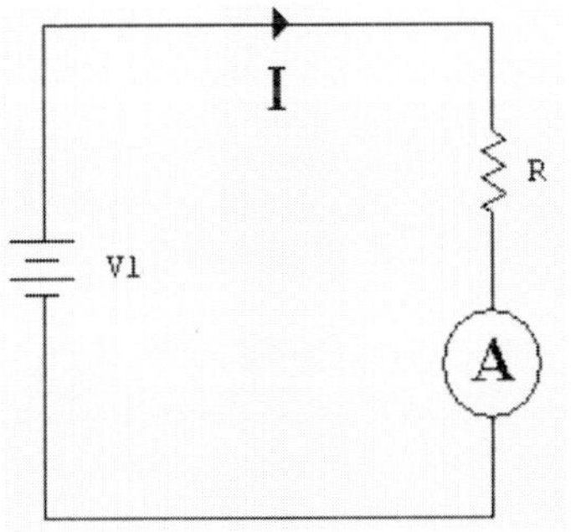

Figure 3-1

There are two quantities that affect current flow: voltage and resistance. When resistance is constant, voltage and current are directly related. When voltage is constant, current and resistance are inversely related.

When voltage and resistance are known, current may be predicted using a relationship known as Ohm's law. Ohm's law states: one volt is required to cause a one-amp current flow through one ohm of resistance. In arithmetic form, this is:

$$I=V/R$$

Learning Objectives

Upon completion of this lab, the student will be able to:

1. Configure a circuit using eTCB schematic drawing and eTCB board.
2. Predict current, given voltage and resistance.
3. Use a digital multimeter (DMM) to measure current through a circuit.

Pre-Lab Procedure

1. Use the Ohm's law equation to predict the current I, for different values of voltage V1. Record your results in Table 3-1 (column – Predicted Current). Let the resistance R equal to 470 Ω.

2. Use the Ohm's law equation to predict the current I, for different values of voltage V1. Record your results in Table 3-1 (column – Predicted Current). Let the resistance R equal to 820 Ω.

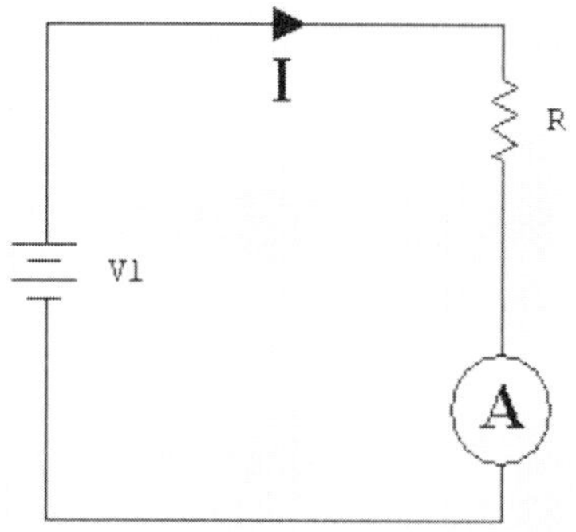

Figure 3-2

Lab Procedure

Before proceeding further, make sure that your eTCB is working correctly. Using eTCB Diagnostic VI, perform a self-diagnostic of eTCB circuitry.

Turn the switch S1 and S2 OFF in preparation for the lab.

Setup

Step #1 Using the eTCB Configuration VI, (DC module), first reset all relays and then configure the circuit in Figure 3-3.

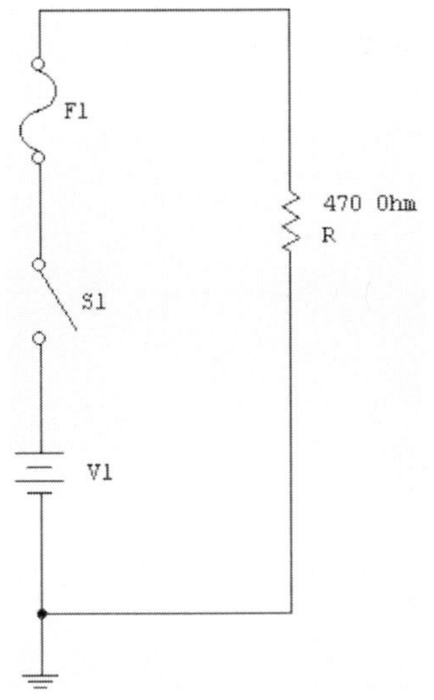

Figure 3-3

You will be asked to measure a DC current in this experiment, and as part of the DMM setup, you will need to "Null" the ammeter. This will minimize the error in the DC current measurement. The "Nulling" procedure is very simple and needs to be performed only once per experiment. The procedure to "Null" the ammeter is as follows:

- Remove the test leads from the DMM terminals labeled "CURRENT".
- Open the Digital Multimeter VI and select DC Current measurement function.
- Click on the "Null" button.

Step #2 Using the Variable Power Supplies VI, set voltage source V1 to ***2.0V.*** Keep in mind that this setting is just a good indication of what the voltage of V1 is set to. Variable power supplies often need to be adjusted (tuned) in order to set their output voltage precisely. Keep the Variable Power Supplies VI ***open***; if you close it, all the variable voltage supplies will reset to 0V.

Step #3 Physically turn switch S1 on the eTCB board "ON". This will complete your series circuit. To verify that the switch is ON (closed) and that the Fuse is not blown, use the Digital Multimeter VI to measure voltage at TP1 with respect to ground. This step of the experiment is designed to verify the output of the voltage source V1. Refer to Figure 3-4 for help in placing the voltmeter.

Note: If you have a second pair of DMM leads, this would be an ideal time to use them. Keep the voltmeter test leads connected on the eTCB board and inserted in the NI ELVIS terminals for the duration of this lab. Use the second pair of test leads to connect the ammeter. If you don't have a second pair, you will need to switch the connection back and forth between measuring voltage and current.

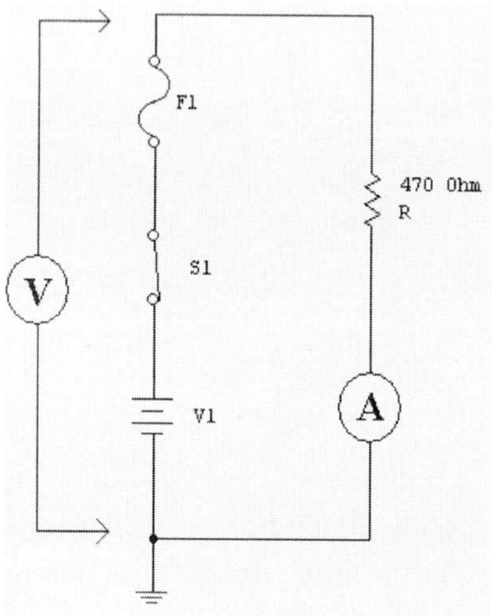

Figure 3-4

Discovery

Step #4 Using the Digital Multimeter VI, measure current in the circuit. Record this reading in Table 3-1 (column – Measured Current). To perform this measurement, you need to insert the ammeter in series with the resistor R. Removing the jumper J2 creates a break in the circuit where the ammeter can be inserted.

Note: Use the second pair of DMM leads to insert the ammeter into the circuit. Keep the voltmeter test leads just like they were connected in the previous step.

Step #5 Using the Variable Power Supplies VI, set voltage source V1 to ***4.0V.***
Using the Digital Multimeter VI, measure voltage at TP1 with respect to ground. Adjust the output of the variable power supplies, if necessary.

Step #6 Using the Digital Multimeter VI, measure current in the circuit. Record this reading in Table 3-1 (column – Measured Current).

Step #7 Using the Variable Power Supplies VI, set voltage source V1 to ***6.0V.***
Using the Digital Multimeter VI, measure voltage at TP1 with respect to ground. Adjust the output of the variable power supplies, if necessary.

Step #8 Using the Digital Multimeter VI, measure current in the circuit. Record this reading in Table 3-1 (column – Measured Current).

Step #9 Using the Variable Power Supplies VI, set voltage source V1 to ***8.0V.***
Using the Digital Multimeter VI, measure voltage at TP1 with respect to ground. Adjust the output of the variable power supplies, if necessary.

Step #10 Using the Digital Multimeter VI, measure current in the circuit. Record this reading in Table 3-1 (column – Measured Current).

Step #11 Using the Variable Power Supplies VI, set voltage source V1 to ***10.0V.***
Using the Digital Multimeter VI, measure voltage at TP1 with respect to ground. Adjust the output of the variable power supplies, if necessary.

Step #12 Using the Digital Multimeter VI, measure current in the circuit. Record this reading in Table 3-1 (column – Measured Current).

Note: Now we are going to examine the effects of different resistance on current flow. We are going to use the same circuit diagram, but rather than using 470 Ω, we are going to use 820 Ω resistor. The eTCB is designed with two resistors in each of the circuit branches. In this experiment, it is the insertion of the ammeter that determines which of the two resistors will be included in the circuit. Therefore, to include 820 Ω resistor, the ammeter will need to be inserted between the 820 Ω resistor and the GND node. (See Figure 3.5)

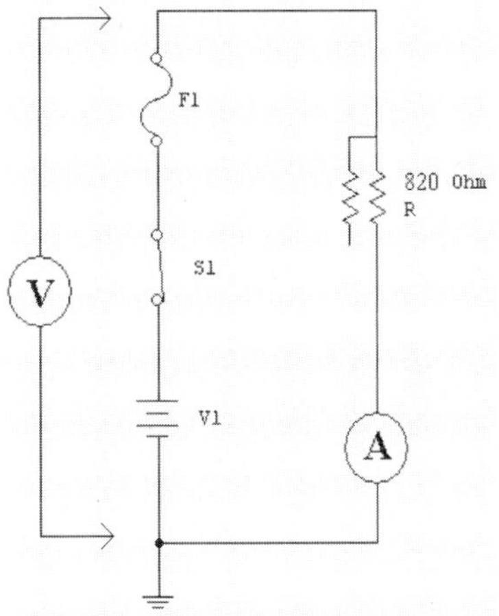

Figure 3-5

Step #13 Using the Variable Power Supplies VI, set voltage source V1 to ***2.0V.***
Using the Digital Multimeter VI, measure voltage at TP1 with respect to ground. Adjust the output of the variable power supplies, if necessary.

Step #14 Using the Digital Multimeter VI, measure current in the circuit. Record this reading in Table 3-1 (column – Measured Current). To perform this measurement, you need to insert the ammeter in series with the resistor R. Removing the jumper J2 creates a break in the circuit where the ammeter can be inserted.

Step #15 Using the Variable Power Supplies VI, set voltage source V1 to ***4.0V.***
Using the Digital Multimeter VI, measure voltage at TP1 with respect to ground. Adjust the output of the variable power supplies, if necessary.

Step #16 Using the Digital Multimeter VI, measure current in the circuit. Record this reading in Table 3-1 (column – Measured Current).

Step #17 Using the Variable Power Supplies VI, set voltage source V1 to ***6.0V.***
Using the Digital Multimeter VI, measure voltage at TP1 with respect to ground.Adjust the output of the variable power supplies, if necessary.

Step #18 Using the Digital Multimeter VI, measure current in the circuit. Record this reading in Table 3-1 (column – Measured Current).

Step #19 Using the Variable Power Supplies VI, set voltage source V1 to ***8.0V***
Using the Digital Multimeter VI, measure voltage at TP1 with respect to ground. Adjust the output of the Variable Power Supplies if necessary.

Step #20 Using the Digital Multimeter VI, measure current in the circuit. Record this reading in Table 3-1. (Column – Measured Current).

Step #21 Using the Variable Power Supplies VI, set voltage source V1 to ***10.0V***
Using the Digital Multimeter VI, measure voltage at TP1 with respect to ground. Adjust the output of the Variable Power Supplies if necessary.

Step #22 Using the Digital Multimeter VI, measure current in the circuit. Record this reading in Table 3-1 (column – Measured Current).

This concludes the procedure for this experiment.

Table 3-1

Voltage V1	Predicted Current	Measured Current
R=470 Ω		
2.0V		
4.0V		
6.0V		
8.0V		
10.0V		
R=820 Ω		
2.0V		
4.0V		
6.0V		
8.0V		
10.0V		

Questions

1. How is the ammeter connected to the circuit in order to measure current?

2. What is the resistance of an ideal ammeter?

3. Perform the error calculation for all entries in Table 3-1. Comment on whether the errors are acceptable.

4. On the same graph, plot the Current vs. Voltage. Use the measured current values from Table 3-1.

5. If voltage across a resistor is increased, what effect does it have on the current through that resistor?

6. If resistance is increased while voltage is kept constant, what effect does it have on the current through that resistor?

7. What is the relationship of the slope of the Current vs. Voltage curve to Conductance and Resistance?

Bonus:

Consider an electric heater (space heater), one commonly used in a household. The heat generated in that type of appliance is a direct result of a current flow through a resistor. The resistor (or in this case, the heating element) is designed so that it can withstand the heat it generates. How would you propose we increase the heat generated in this electric heater?

4

Voltage

Problem Statement

Voltage refers to accumulation of electrical charge. By understanding how to control voltage, one can design, analyze and troubleshoot electrical circuits. Consider a scenario where there is a need to determine a current flow through a resistor on an assembled printed circuit board (PCB). As you know, in order to measure a current flow with an ammeter, a break in a circuit needs to be made, and the ammeter is then inserted to complete the original circuit. In our scenario, however, there are not easy ways to create a break in the circuit, as it is assembled on a printed circuit board. In this scenario, we can use our understanding of the voltage, current, and resistance relationship and determine the current flow through a resistor by measuring voltage developed across it. Determining current flow by measuring voltage across a known resistor is a common method used in situations where ammeters cannot easily be inserted in the path of a current flow.

Concept

Voltage (V or E) is a force known as electrical pressure. It is the force that causes current to overcome resistance and flow in a circuit. This pressure is referred to as electromotive force (emf). There are several types of voltage sources: 1) batteries, 2) generators, 3) solar cells, 4) piezo electric crystals, 5) thermocouples, and 6) photo voltaic cells. The unit of measurement for voltage is the volt (V). Voltage is measured with a voltmeter. Voltmeters are placed in parallel with the component in order to measure voltage, as shown in Figure 4-1.Voltmeters ideally represent an open or infinite resistance when used to measure voltage in a circuit. With infinite resistance, zero current flows through the meter, so the meter does not affect the operation of the circuit.

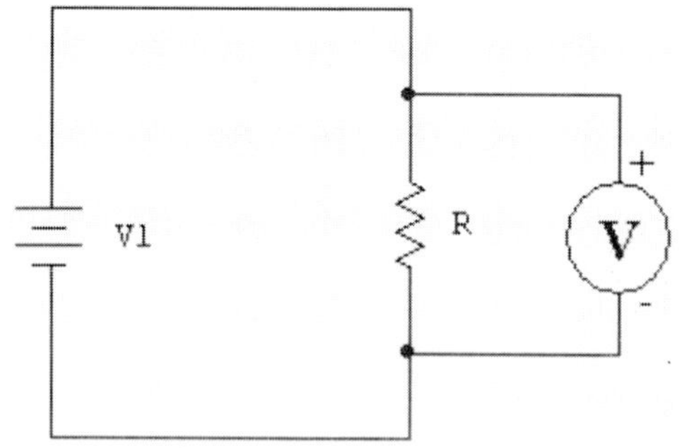

Figure 4-1

When current and resistance are known, voltage may be determined using Ohm's law. Ohm's law states: one volt is required to cause a one amp current flow through one ohm of resistance. In arithmetic form, it is:

$$V=IR$$

Looking at Ohm's law, it can be seen that the larger the resistance, the greater the voltage necessary to cause the desired current to flow.

Learning Objectives

Upon completion of this lab, the student will be able to:

1. Configure a circuit using eTCB schematic drawing and eTCB board.
2. Predict voltage, given current and resistance.
3. Use a digital multimeter (DMM) to measure voltage.

Pre-Lab Procedure

1. Use the Ohm's law equation to predict the voltage V across the resistor R, for different values of current. Record your results in Table 4-1 (column – Predicted Voltage). Let the resistance R equal to 470 Ω. (See Figure 4.2)

2. Use the Ohm's law equation to predict the voltage V across the resistor R, for different values of current. Record your results in Table 4-1 (column – Predicted Voltage). Let the resistance R equal to 820 Ω.

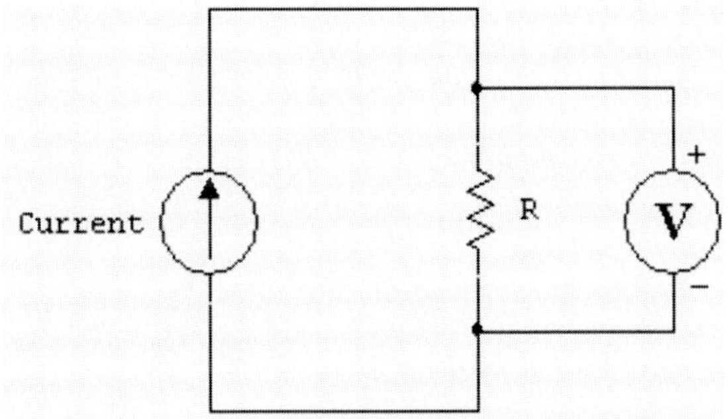

Figure 4-2

Lab Procedure

Before proceeding further, make sure that your eTCB is working correctly. Using eTCB Diagnostic VI, perform a self-diagnostic of eTCB circuitry.

Turn the switch S1 and S2 OFF in preparation for the lab.

Setup

Step #1 Using the eTCB Configuration VI, (DC module), first reset all relays and then configure the circuit, as shown in Figure 4-3.

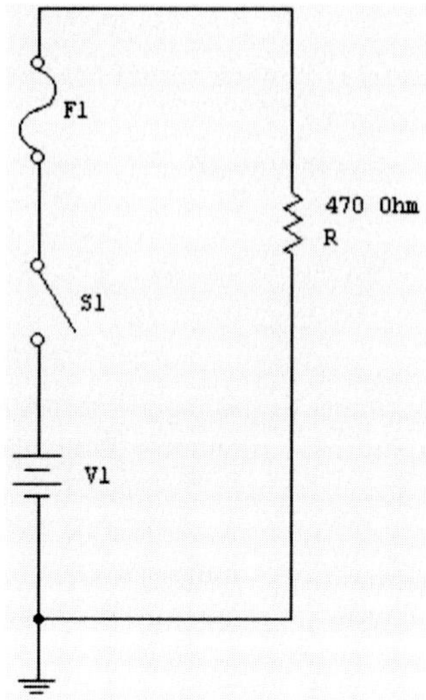

Figure 4-3

You will be asked to measure a DC current in this experiment, and as part of the DMM setup you need to "Null" the ammeter. This will minimize the error in DC current measurement. The "Nulling" procedure is very simple and needs to be performed only once per experiment. The procedure to "Null" the ammeter is as follows:

- Remove the test leads from the DMM terminals labeled "CURRENT".
- Open the Digital Multimeter VI and select DC Current measurement function.
- Click on the "Null" button.

Step #2 Using the Variable Power Supplies VI, set voltage source V1 to ***1.5V.*** Keep in mind that this setting is just a good indication of what the voltage of V1 is set to. Variable power supplies often need to be adjusted (tuned) in order to set their output voltage precisely. Keep the Variable Power Supplies VI ***open***; if you close it, all the variable voltage supplies will reset to 0V.

Step #3 Physically turn switch S1 on the eTCB board "ON". This will complete your series circuit. To verify that the switch is "ON" (closed) and that the fuse is not blown, use the Digital Multimeter VI to measure voltage at TP1 with respect to ground. This step of the experiment is designed to verify the output of the voltage source V1.

Discovery

Note: *We do not have a true current source available to us for this experiment. To emulate a variable current source, we are going to vary the output voltage V1; this will in turn cause different current flow, depending on the voltage value.*

Step #4 Using the Digital Multimeter VI, measure current in the circuit while adjusting the Variable Power Supply output V1 to obtain a current flow of **4mA.** To perform this measurement, you need to insert the ammeter in series with the resistor R. Removing the jumper **J2** creates a break in the circuit where the ammeter can be inserted. Refer to Figure 4-3 for information on meter connections.

Note: *If you have a second pair of DMM leads, this would be an ideal time to use them. Keep the ammeter test leads connected on the eTCB board and inserted in the NI ELVIS terminals for the duration of this lab. Use the second pair of test leads to connect the voltmeter. If you don't have a second pai,r you will need to switch the connection back and forth between measuring current and voltage.*

Step #5 Using the Digital Multimeter VI, measure voltage across resistor R. Record this reading in Table 4-1 (column – Measured Voltage). Refer to Figure 4-3 for information on meter connections.

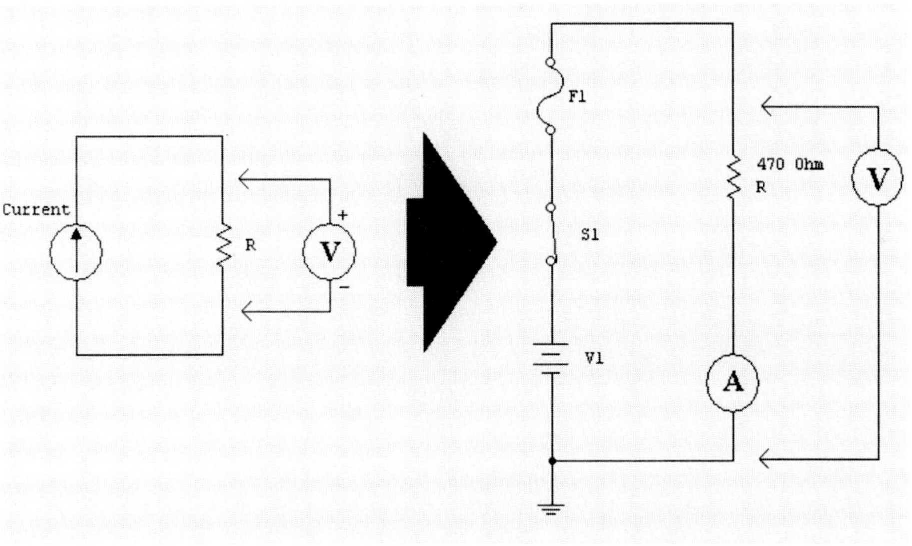

Figure 4-3

Step #6 Using the Digital Multimeter VI, measure current in the circuit while adjusting the Variable Power Supply output V1 to obtain a current flow of **8mA.**

Step #7 Using the Digital Multimeter VI, measure voltage across resistor R. Record this reading in Table 4-1 (column – Measured Voltage).

Step #8 Using the Digital Multimeter VI, measure current in the circuit while adjusting the Variable Power Supply output V1 to obtain a current flow of **12mA.**

Step #9 Using the Digital Multimeter VI, measure voltage across resistor R. Record this reading in Table 4-1 (column – Measured Voltage).

Step #10 Using the Digital Multimeter VI, measure current in the circuit while adjusting the Variable Power Supply output V1 to obtain a current flow of **16mA.**

Step #11 Using the Digital Multimeter VI, measure voltage across resistor R. Record this reading in Table 4-1 (column – Measured Voltage).

Step #12 Using the Digital Multimeter VI, measure current in the circuit while adjusting the Variable Power Supply output V1 to obtain a current flow of **20mA.**

Step #13 Using the Digital Multimeter VI, measure voltage across resistor R. Record this reading in Table 4-1 (column – Measured Voltage).

Note: We are going to examine the effects of different resistance on voltage developed across the resistor. We are going to use the same circuit diagram, but rather than using 470 Ω, we are going to use 820 Ω resistor. The eTCB is designed with two resistors in each of the circuit branches. In this experiment, it is the insertion of the ammeter that determines which one of the two resistors will be included in the circuit (Figure 4-4). Therefore, to include 820 Ω resistor, the ammeter will need to be inserted between the 820 Ω resistor and the GND node.

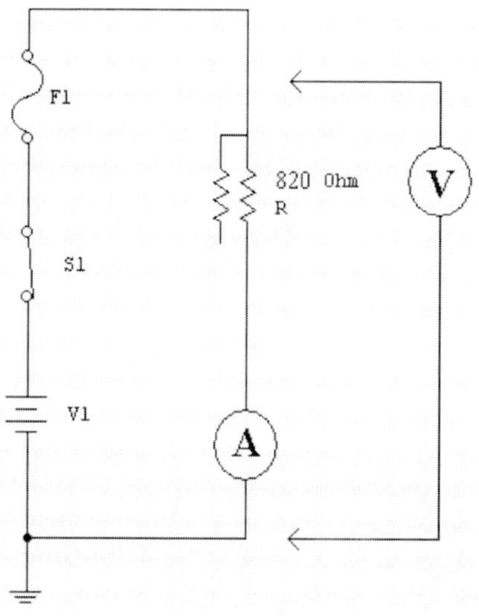

Figure 4-4

Step #14 Using the Digital Multimeter VI, measure current in the circuit while adjusting the Variable Power Supply output V1 to obtain a current flow of **2mA.** To perform this measurement, you need to insert the ammeter in series with the resistor R. Removing the jumper J2 creates a break in the circuit where the ammeter can be inserted.

Step #15 Using the Digital Multimeter VI, measure voltage across resistor R. Record this reading in Table 4-1 (column – Measured Voltage).

Step #16 Using the Digital Multimeter VI, measure current in the circuit while adjusting the Variable Power Supply output V1 to obtain a current flow of **4mA.**

Step #17 Using the Digital Multimeter VI, measure voltage across resistor R. Record this reading in Table 4-1 (column – Measured Voltage).

Using the Digital Multimeter VI, measure voltage at TP1 with respect to ground. Adjust the output of the variable power supplies, if necessary.

Step #18 Using the Digital Multimeter VI, measure current in the circuit while adjusting the Variable Power Supply output V1 to obtain a current flow of **6mA.**

Step #19 Using the Digital Multimeter VI, measure voltage across resistor R. Record this reading in Table 4-1 (column – Measured Voltage).

Step #20 Using the Digital Multimeter VI, measure current in the circuit while adjusting the Variable Power Supply output V1 to obtain a current flow of **8mA.**

Step #21 Using the Digital Multimeter VI, measure voltage across resistor R. Record this reading in Table 4-1 (column – Measured Voltage).

Step #22 Using the Digital Multimeter VI, measure current in the circuit while adjusting the Variable Power Supply output V1 to obtain a current flow of **10mA.**

Step #23 Using the Digital Multimeter VI, measure voltage across resistor R. Record this reading in Table 4-1 (column – Measured Voltage).

This concludes the procedure for this experiment.

Table 4-1

Current	*Predicted Voltage*	*Measured Voltage*
R=470 Ω		
4mA		
8mA		
12mA		
16mA		
20mA		
R=820 Ω		
2mA		
4mA		
6mA		
8mA		
10mA		

Questions

1. How is the voltmeter connected to the circuit in order to measure voltage?
2. What is the resistance of an ideal voltmeter?
3. Perform the error calculation for all entries in Table 4-1. Comment on whether the errors are acceptable.
4. On the same graph, plot the Voltage vs. Current. Use the measured voltage values from Table 4-1.
5. If current through a resistor is increased, what effect does it have on the voltage across that resistor?
6. If resistance is increased while current is kept constant, what effect does it have on the resistor voltage?
7. What is the relationship of the slope of the Voltage vs. Current curve to Conductance and Resistance?

Bonus:

8. When determining the polarity of voltage across a resistive load in a circuit, the resistor end where the electron current enters is labeled the (negative or positive) end?

5

Resistance

Problem Statement

The concept of voltage and current was introduced in the previous experiments. You have determined that current relates to the movement of free electrons (charge). In a conductor, the charge moves due to the voltage of a power source. The opposition to charge movement is called resistance. In a process similar to friction, a moving charge gives up some of its energy in a form of heat or light. By understanding electrical resistance, you can better understand how the light bulb and the common household space heater works.

Let us consider an application of the resistance concept in a common hair dryer. The hair dryer is connected to a wall plug that is its power source. The blower motor is rated at a fraction of the voltage from the wall plug. This is to make it cost effective. Let's say the motor is rated at 12V AC. Our challenge is to construct the circuit that can be powered with 120V AC (wall plug) and use the 12V AC motor. If we were to connect the blower motor directly to the wall plug, it would draw too much current and burn out. This is because blower motor has low resistance. We would also exceed its voltage rating. This is where a resistor placed between the wall plug and the blower motor can reduce the current drawn from the wall plug (see Figure 5-1). This resistor could also serve as a heater to generate the required heat for the hair dryer. By understanding what the resistance does in electrical circuits, we can determine how to build a cost-effective and functional hair dryer and more.

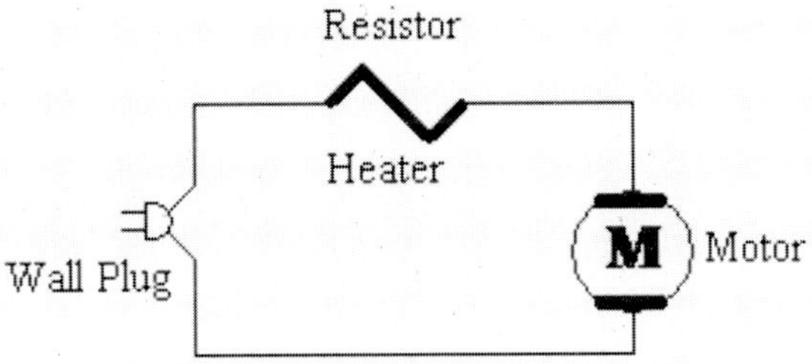

Figure 5-1

Concept

Resistance is the opposition to the flow of current. Resistance is measured in ohms. The symbol for ohms is omega, Ω. The greater the opposition to the flow of current a material or device has, the more ohms of resistance are present in the circuit.

Resistance is measured with an ohmmeter, which represents a voltage source to the circuit. **Never insert an ohmmeter in a circuit where power is applied!** Circuits or paths with little or no resistance are referred to as short circuits. Circuits with infinite resistance are referred as open circuits.

Resistance and current are inversely related. If the voltage is kept constant, then the greater the resistance, the lower the current. The opposite is also true: the lower the resistance, the greater the current. This relationship is demonstrated in the Ohm's law equation for resistance, as shown:

$$\mathbf{R=V/I}$$

Learning Objectives

Upon completion of this lab, the student will be able to:

1. Configure a circuit using eTCB schematic drawing and eTCB board.
2. Determine the resistance and the tolerance of a color-coded resistor.
3. Predict the resistance, given current and voltage.
4. Use a digital multimeter (DMM) to measure resistance.

Pre-Lab Procedure

1. Use the Ohm's law equation to predict the resistance of the resistor R, for different values of current, and voltage. Record your results in Table 5-1.

Table 5-1

Voltage across the Resistor	***Current through the Resistor***	***Resistance of the Resistor***
4.0V	8.51mA	
8.0V	17.02mA	
4.0V	4.878mA	
8.0V	9.756mA	

2. Use the resistor color code to determine the resistance value and the tolerance of the resistors described in Table 5-2.

Table 5-2

First Band	***Second Band***	***Third Band***	***Forth Band***	***Resistance Value***	***Tolerance***
Yellow	*Purple*	*Brown*	*Gold*		
Gray	*Red*	*Brown*	*Gold*		

Lab Procedure

Before proceeding further, make sure that your eTCB is working correctly. Using eTCB Diagnostic VI, perform a self-diagnostic of eTCB circuitry.

Turn the switch S1 and S2 OFF in preparation for the lab.

Setup

Step #1 Using the eTCB Configuration VI, (DC Module), first reset all relays and then configure the circuit as shown in Figure 5-2.

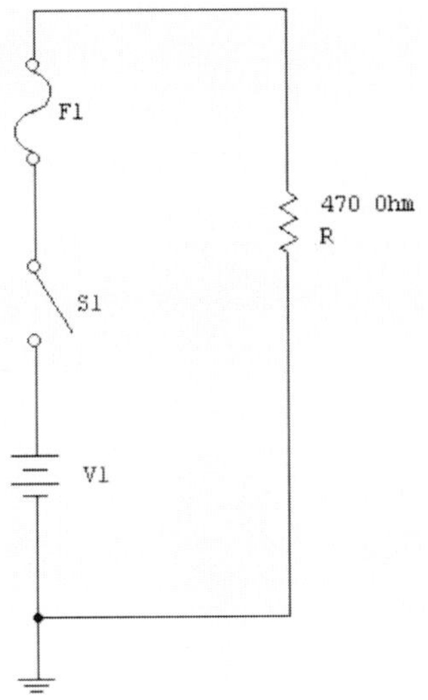

Figure 5-2

You will be asked to measure a DC current in this experiment, and as part of the DMM setup, you need to "Null" the ammeter. This will minimize the error in DC current measurement. The "Nulling" procedure is very simple and needs to be performed only once per experiment. The procedure to "Null" the ammeter is as follows:

- Remove the test leads from the DMM terminals labeled "CURRENT".
- Open the Digital Multimeter VI and select DC Current measurement function.
- Click on the "Null" button.

Step #2 Using the Variable Power Supplies VI, set voltage source V1 to ***2.0.V*** Keep in mind that this setting is just a good indication of what the voltage of V1 is set to. Variable power supplies often need to be adjusted (tuned) in order to set their output voltage precisely. Keep the Variable Power Supplies VI ***open***; if you close it, all the variable voltage supplies will reset to 0V.

Step #3 Physically turn switch S1 on the eTCB board "ON". This will complete your series circuit. To verify that the switch is ON (closed) and that the vuse is not blown, use the Digital Multimeter VI to measure voltage at TP1 with respect to ground. This step of the experiment is designed to verify the output of the voltage source V1. Refer to Figure 5-3 for help in placing the voltmeter.

Note: If you have a second pair of DMM leads, this would be an ideal time to use them. Keep the voltmeter test leads connected on the eTCB board and inserted in the NI ELVIS terminals for the duration of this lab. Use the second pair of test leads to connect the ammeter. If you don't have a second pair you will need to switch the connection back and forth between measuring voltage and current.

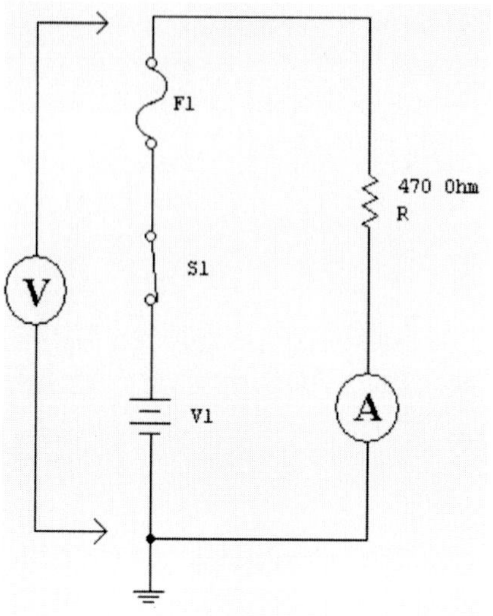

Figure 5-3

Discovery

Step #4 Using the Digital Multimeter VI, measure current in the circuit. Record this reading in Table 5-5 (column– Measured Current). To perform this measurement you need to insert the ammeter in series with the resistor R. Removing the jumper J2 creates a break in the circuit where the ammeter can be inserted.

Note: Use the second pair of DMM leads to insert the ammeter into the circuit. Keep the voltmeter test leads just like thy were connected in the previous step.

Step #5 Using the Variable Power Supplies VI, set voltage source V1 to ***6.0V.***
Using the Digital Multimeter VI, measure voltage at TP1 with respect to ground. Adjust the output of the variable power supplies, if necessary.

Step #6 Using the Digital Multimeter VI, measure current in the circuit. Record this reading in Table 5-5 (column– Measured Current).

Step #7 Using the Variable Power Supplies VI, set voltage source V1 to ***10.0V.***
Using the Digital Multimeter VI, measure voltage at TP1 with respect to ground. Adjust the output of the variable power supplies, if necessary.

Step #8 Using the Digital Multimeter VI, measure current in the circuit. Record this reading in Table 5-5 (column– Measured Current).

Note: *Now we are going to examine the effects of different resistance on circuit performance. We are going to use the same circuit diagram, but rather than using 470 Ω, we are going to use 820 Ω resistor. The eTCB is designed with two resistors in each of the circuit branches. In this experiment, it is the insertion of the ammeter that determines which of the two resistors will be included in the circuit (Figure 5-4). Therefore, to include 820 Ω resistor, the ammeter will need to be inserted between the 820 Ω resistor and the GND node.*

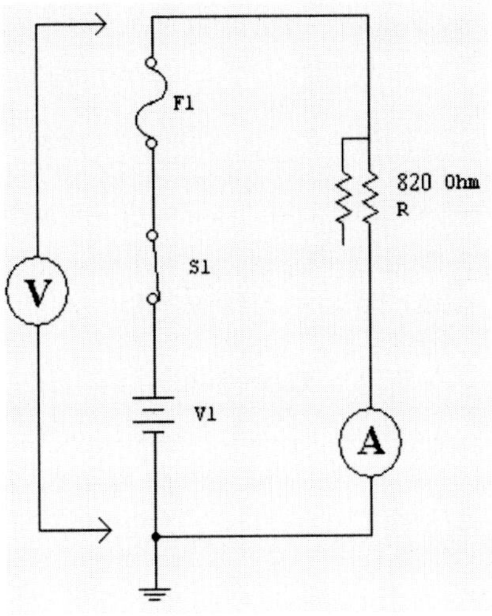

Figure 5-4

Step #9 Using the Variable Power Supplies VI, set voltage source V1 to ***2.0V.***
Using the Digital Multimeter VI, measure voltage at TP1 with respect to ground. Adjust the output of the variable power supplies, if necessary.

Step #10 Using the Digital Multimeter VI, measure current in the circuit. Record this reading in Table 5-5 (column– Measured Current).

Step #11 Using the Variable Power Supplies VI, set voltage source V1 to ***6.0V.***
Using the Digital Multimeter VI, measure voltage at TP1 with respect to ground. Adjust the output of the variable power supplies, if necessary.

Step #12 Using the Digital Multimeter VI, measure current in the circuit. Record this reading in Table 5-5 (column– Measured Current).

Step #13 Using the Variable Power Supplies VI, set voltage source V1 to ***10.0V.***
Using the Digital Multimeter VI, measure voltage at TP1 with respect to ground. Adjust the output of the variable power supplies, if necessary.

Step #14 Using the Digital Multimeter VI, measure current in the circuit. Record this reading in Table 5-5 (column– Measured Current).

Note: *You are now going to measure the resistance using the digital multimeter (DMM)*

Step #15 Using the eTCB Configuration VI, (DC module), reconfigure the circuit so that no voltage supply is connected to any of the electrical components available on the eTCB board. Also make sure that the 470 Ω resistor and 820 Ω resistor pair are electrically isolated. (Capture configuration data to verify its accuracy.) **Resistance measurements can be performed only when all power sources are removed.**

Step #16 Using the Digital Multimeter VI, measure resistance of the 470 Ω resistor on the eTCB board. Record this reading below in Table 5-3.

Table 5-3

DMM Measured Resistance	

Step #17 Using the Digital Multimeter VI, measure resistance of the 820 Ω resistor on the eTCB board. Record this reading below in Table 5-4.

Table 5-4

DMM Measured Resistance	

This concludes the procedure for this experiment.

Table 5-5

470 Ω Resistor		
Voltage V1	***Measured Current***	***Measured Resistance***
2.0V		*
6.0V		*
10.0V		*
820 Ω Resistor		
Voltage V1	***Measured Current***	***Measured Resistance***
2.0V		*
6.0V		*
10.0V		*

	* Measured Resistance refers to the resistance calculated using measured values of voltage and current.

Questions

1. Perform the error calculation for all entries in Table 5-5. Comment on whether the errors are acceptable.

2. On the same graph, plot the Voltage vs. Current. Use the measured voltage values from Table 5-5. How does the slope of these curves relate to the resistance.

3. When examining a circuit as shown in Figure 5-3, you notice that the voltage across the resistor is within the acceptable range, but the current through it is too high. What does this suggest about the resistance value of the resistor?

Bonus:

When examining a circuit as shown in Figure 5-3, you notice that the current through the resistor is too high, and the resistor measures within the acceptable range (within tolerance). What could be the problem.

6

Power

Problem Statement

Household electric heaters and light bulbs are rated in watts (W), and the motors are rated in horsepower (or watts): both are the units of power. The greater the power rating on the heater or the light bulb, the more heat or light is produced in a given time. Every light fixture comes with a label stating the maximum wattage for the light bulb that can be used, but how is this maximum power rating determined? Understanding power dissipation concepts helps us construct circuits that perform close to their rated capacity but not exceeding it.

Concept

Power (P) is the rate at which work is done and is measured in watts (W). Some devices manifest their work in the form of heat. Resistors are an example of this. In terms of the physical size, the larger a resistor is, the more heat it can handle and the greater the power it can dissipate. Incandescent lamps manifest power in the form of light: the higher the wattage, the brighter the lamp. Motors are rated in horsepower (hp). One horsepower is equal to 746 watts.

When power is transferred from an energy source to a load, some energy is lost to heat. Maximum power is transferred from a source to a load when the resistance of the source equals the resistance of the load. This is known as the maximum power transfer theorem. The efficiency of a system is represented in the form of the percentage of power transfer and is determined by:

Efficiency %=(power out/power in)*100%

There is a relationship between voltage, current, and power; it is known as the power equation. The power equation may be stated three ways:

$$P=V*I \qquad V=P/I \qquad I=P/V$$

From these equations it can be seen that power is directly related to voltage and current. It may also be demonstrated that when power is constant, voltage and current are inversely related. Power may also be calculated three ways:

$$P=I^2 R \qquad P=V^2/R \qquad P=VI$$

Learning Objectives

Upon completion of this lab, the student will be able to:

1. Configure a circuit using eTCB schematic drawing and eTCB board.
2. Determine the power dissipation in an electrical load given any two electrical parameters: voltage, current, or resistance.

Pre-Lab Procedure

1. Given voltage across and the resistance of a 470 Ω resistor, use all three different power equation to predict the power dissipated in the resistor R. Perform this calculation for different values of the resistor voltage. Record your results in Table 6-1.

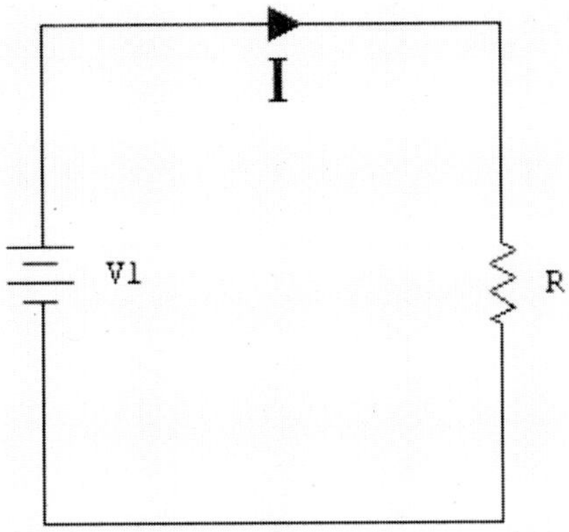

Figure 6-1

2. Repeat Pre-Lab Procedure 1 with an 820 Ω resistor. Record your results in Table 6-2.

Lab Procedure

Before proceeding further, make sure that your eTCB is working correctly. Using eTCB Diagnostic VI, perform a self-diagnostic of eTCB circuitry.

Turn the switch S1 and S2 OFF in preparation for the lab.

Setup

Step #1 Using the eTCB Configuration VI, (DC module), first reset all relays and then configure the circuit in Figure 6-2.

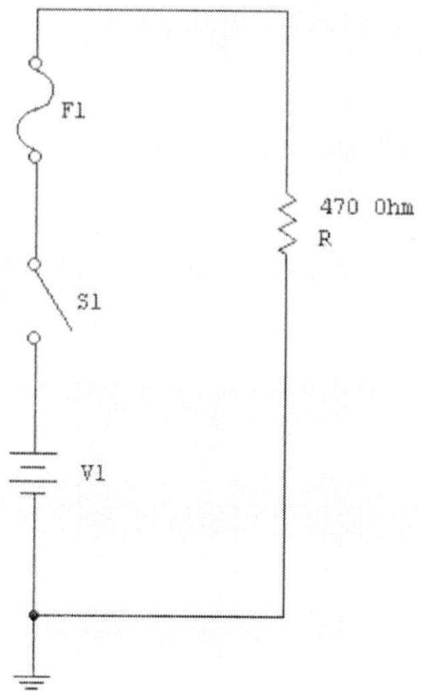

Figure 6-2

You will be asked to measure a DC current in this experiment, and as part of the DMM setup you need to "Null" the ammeter. This will minimize the error in DC current measurement. The "Nulling" procedure is very simple and needs to be performed only once per experiment. The procedure to "Null" the ammeter is as follows:

- Remove the test leads from the DMM terminals labeled "CURRENT".
- Open the Digital Multimeter VI and select DC Current measurement function.
- Click on the "Null" button.

Step #2 Using the Variable Power Supplies VI, set voltage source V1 to ***3.0V.*** Keep in mind that this setting is just a good indication of what the voltage of V1 is set to. Variable power supplies often need to be adjusted (tuned) in order to set their output voltage precisely. Keep the Variable Power Supplies VI ***open***; if you close it, all the variable voltage supplies will reset to 0V.

Step #3 Physically turn switch S1 on the eTCB board "ON". This will complete your series circuit. To verify that the switch is ON (closed) and that the fuse is not blown, use the Digital Multimeter VI to measure voltage at TP1 with respect to ground. Adjust the output of the Variable power supplies, if necessary. Refer to Figure 6-3 for help in placing the voltmeter.

Note: If you have a second pair of DMM leads, this would be an ideal time to use them. Keep the voltmeter test leads connected on the eTCB board and inserted in the NI ELVIS terminals for the duration of this lab. Use the second pair of test leads to connect the ammeter. If you don't have a second pair you will need to switch the connection back and forth between measuring voltage and current.

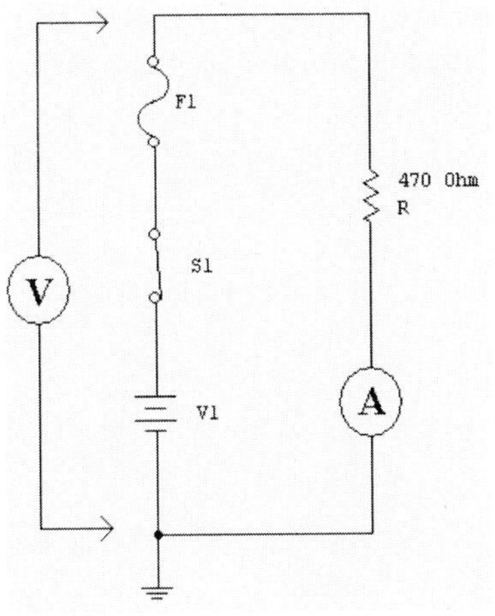

Figure 6-3

Discovery

Step #4 Using the Digital Multimeter VI, measure current in the circuit. Record this reading in Table 6-1 (column – Measured). To perform this measurement, you need to insert the ammeter in series with the resistor R. Removing the jumper J2 creates a break in the circuit where the ammeter can be inserted.

Carefully, touch the resistor R to get a sense of how warm it feels.

Note: Use the second pair of DMM leads to insert the ammeter into the circuit. Keep the voltmeter test leads just like they were connected in the previous step.

Step #5 Using the Variable Power Supplies VI, set voltage source V1 to ***7.0V.***
Using the Digital Multimeter VI, measure voltage at TP1 with respect to ground. Adjust the output of the variable power supplies, if necessary.

Step #6 Using the Digital Multimeter VI, measure current in the circuit. Record this reading in Table 6-1 (column – Measured).

Carefully, touch the resistor R to get a sense of how warm it feels.

Step #7 Using the Variable Power Supplies VI, set voltage source V1 to ***11.0V.***
Using the Digital Multimeter VI, measure voltage at TP1 with respect to ground. Adjust the output of the variable power supplies, if necessary.

Step #8 Using the Digital Multimeter VI, measure current in the circuit. Record this reading in Table 6-1 (column – Measured).

Carefully, touch the resistor R to get a sense of how warm it feels.

Note: *Now we are going to examine the effects of different resistance on the power dissipation. We are going to use the same circuit diagram, but rather than using 470 Ω we are going to use 820 Ω resistor. The eTCB is designed with two resistors in each of the circuit branches. In this experiment it is the insertion of the ammeter that determines which of the two resistors will be included in the circuit (Figure 6-4). Therefore, to include 820 Ω resistor, the ammeter will need to be inserted between the 820 Ω resistor and the GND node.*

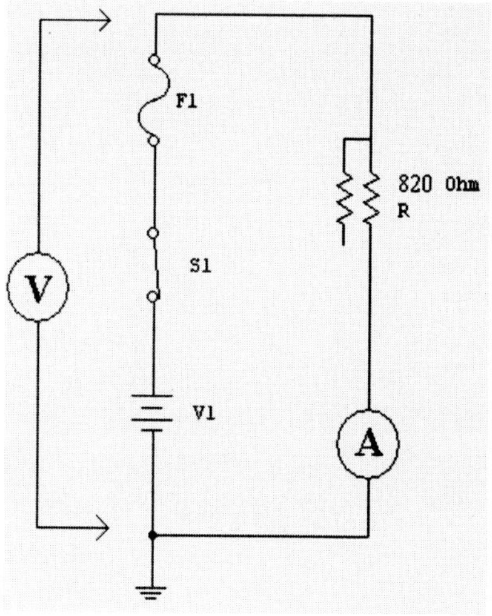

Figure 6-4

Step #9 Using the Variable Power Supplies VI, set voltage source V1 to ***3.0V*** .
Using the Digital Multimeter VI, measure voltage at TP1 with respect to ground. Adjust the output of the variable power supplies, if necessary.

Step #10 Using the Digital Multimeter VI, measure current in the circuit. Record this reading in Table 6-2 (column – Measured).

Carefully, touch the resistor R to get a sense of how warm it feels.

Step #11 Using the Variable Power Supplies VI, set voltage source V1 to ***7.0V.***
Using the Digital Multimeter VI, measure voltage at TP1 with respect to ground. Adjust the output of the variable power supplies, if necessary.

Step #12 Using the Digital Multimeter VI, measure current in the circuit. Record this reading in Table 6-2 (column – Measured).

Carefully, touch the resistor R to get a sense of how warm it feels.

Step #13 Using the Variable Power Supplies VI, set voltage source V1 to ***11.0V*** .
Using the Digital Multimeter VI, measure voltage at TP1 with respect to ground. Adjust the output of the variable power supplies, if necessary.

Step #14 Using the Digital Multimeter VI, measure current in the circuit. Record this reading in Table 6-2 (column – Measured).

Carefully, touch the resistor R to get a sense of how warm it feels.

This concludes the procedure for this experiment.

Table 6-1 (R = 470 Ω)

	Predicted	***Measured***
Voltage	**3.0V**	**3.0V**
Current		
Power (VI)		*
Power (V^2/R)		*
Power (I^2R)		*

Voltage	**7.0V**	**7.0V**
Current		
Power (VI)		*
Power (V^2/R)		*
Power (I^2R)		*

Voltage	**11.0V**	**11.0V**
Current		
Power (VI)		*
Power (V^2/R)		*
Power (I^2R)		*

	*** Calculated based on measured values**

Table 6-2 (R = 820 Ω)

	Predicted	*Measured*
Voltage	**3.0V**	**3.0V**
Current		
Power (VI)		*
Power (V^2/R)		*
Power (I^2R)		*
Voltage	**7.0V**	**7.0V**
Current		
Power (VI)		*
Power (V^2/R)		*
Power (I^2R)		*
Voltage	**11.0V**	**11.0V**
Current		
Power (VI)		*
Power (V^2/R)		*
Power (I^2R)		*
	*** Calculated based on measured values**	

Questions

1. Perform the error calculation for all entries in Table 6-1 and 6-2. Comment on whether the errors are acceptable.

2. On the same graph, plot the Power vs. Voltage. Use the measured values from Table 6-1 and Table 6-2.

3. On the same graph, plot the Power vs. Current. Use the measured values from Table 6-1 and Table 6-2.

4. If voltage is constant and resistance is increased, what effect does it have on the power dissipated in the resistive load?

5. If current is constant and resistance is increased, what effect does it have on the power dissipated in the resistive load?

6. Doubling the current through a resistor has what effect on the power dissipated in the resistive load?

7. Doubling the voltage across a resistor has what effect on the power dissipated in the resistive load?

8. How does power manifest itself in a resistor operating within its rated capacity?

Bonus:

9. A 220 Ω resistor as shown in Figure 6-5 is powered by a variable power supply whose voltage varies from 0-25V. What is the range of power dissipated by the resistor?

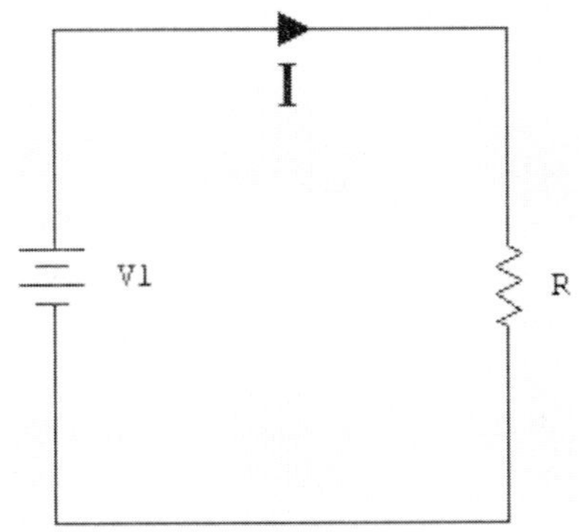

Figure 6-5

7

Resistance in Series

Problem Statement

From time to time, a necessity for a resistor value other than a standard value exists. By understanding how multiple resistors behave when they are connected in series, you will be able to build a circuit that has an equivalent resistance needed by a particular design. For example, let's say we need a resistive load of 472.2 Ω. This resistance value is not a standard one. In this experiment you will practice a concept of combining resistors in series, and you will examine how the equivalent resistance is determined. You will then be able to combine a standard 470Ω resistor and a standard 2.2Ω resistor to build a circuit that meets the initial requirement.

Concept

There is only one path for current to flow in a series circuit. Current must flow through every load in series. The total resistance in a series circuit is equal to the sum of all individual resistances (Figure 7-1). If a single load is shorted out in a series circuit, the total resistance of the circuit will drop. This will cause an increase in the circuit current. The individual voltage drops across all good resistors will increase. Zero volts (no difference of potential) will be measured across the shorted resistor.

$$Req=R1+R2+R3+R4$$

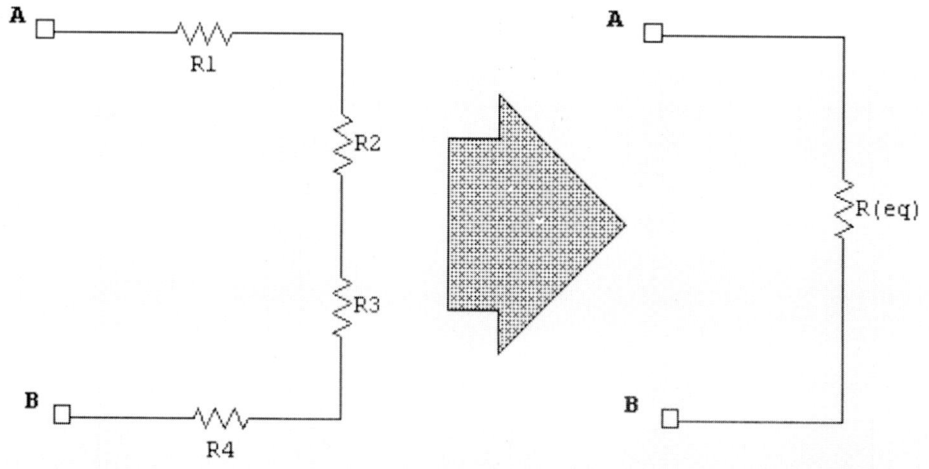

Figure 7-1

If a load in a series circuit is open, the entire circuit is open and the total resistance will increase to infinity. This will cause the circuit current to drop to zero amps. With no current flow there will be no voltage, or difference of potential, across the good resistors. The only measurable voltage drop will be across the open resistor. The source voltage will be measured across the open.

Learning Objectives

Upon completion of this lab the student will be able to:

1. Configure a circuit using eTCB schematic drawing and eTCB board.
2. Predict and measure the resistance in a series circuit.

Pre-Lab Procedure

1. Given the circuit in Figure 7-2 and resistors color code in Table 7-1, determine the resistance of each of the resistors as well as the equivalent resistance seen between points A and B. Record the results in Table 7-1 and Table 7-2.

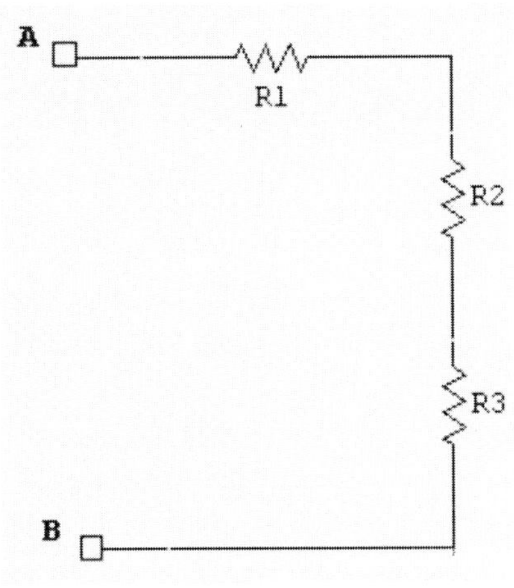

Figure 7-2

Table 7-1

	First Band	***Second Band***	***Third Band***	***Forth Band***	***Rated Resistance***	***Tolerance***
R1	*Red*	*Red*	*Brown*	*Gold*		
R2	*Brown*	*Black*	*Red*	*Gold*		
R3	*Green*	*Blue*	*Brown*	*Gold*		

2. Repeat Pre-Lab Procedure 1 with resistor R1 being shorted.

Lab Procedure

Before proceeding further, make sure that your eTCB is working correctly. Using eTCB Diagnostic VI, perform a self-diagnostic of eTCB circuitry.

Turn the switch S1 and S2 OFF in preparation for the lab.

Setup

Step #1 Using the eTCB Configuration VI, (DC Module), first reset all relays and then configure the circuit in Figure 7-3.

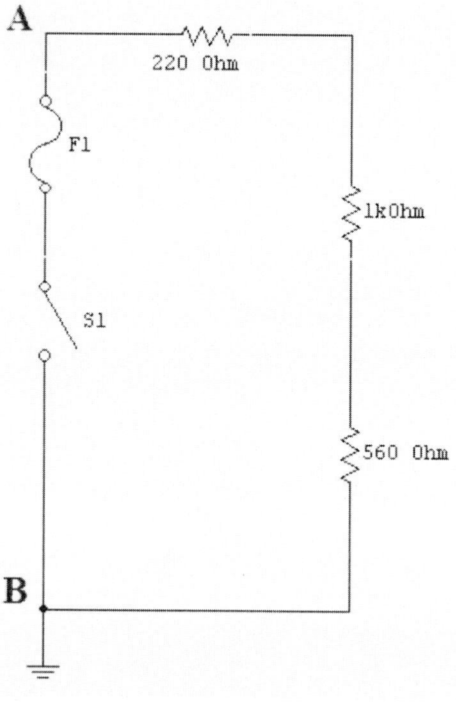

Figure 7-3

Note: *Make sure that switch S1 remains "open" "OFF" and all the power supplies are completely isolated from any components on the eTCB board.*

Discovery

Step #2 Using the Digital Multimeter VI, measure the resistance of **560 Ω** resistor (Figure 704). Record this reading in Table 7-2 (column – Measured).

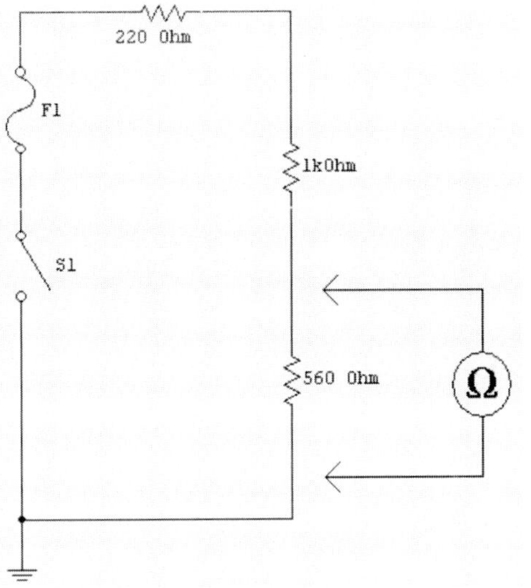

Figure 7-4

Step #3 Using the Digital Multimeter VI, measure the resistance of **1.0 kΩ** resistor. Record this reading in Table 7-2 (column – Measured).

Step #4 Using the Digital Multimeter VI, measure the resistance of **220 Ω** resistor. Record this reading in Table 7-2 (column – Measured).

Step #5 Using the Digital Multimeter VI, measure the equivalent resistance of the series circuit seen from points **A and B.** Record this reading in Table 7-2 (column – Measured).

Step #6 Using the eTCB Configuration VI, (DC module), configure the circuit as shown in Figure 7-5. This circuit has a 220 Ω resistor which is shorted.

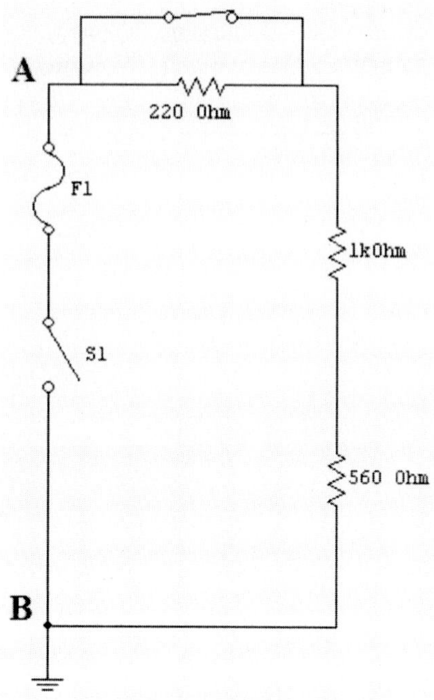

Figure 7-5

Note: *Make sure that switch S1 remains "open" "OFF" and all the power supplies are completely isolated from any components on the eTCB board.*

Step #7 Using the Digital Multimeter VI, measure the equivalent resistance of the series circuit seen from points **A and B.** Record this reading in Table 7-2 (column – Measured).

This concludes the procedure for this experiment.

Table 7-2

	Predicted	***Measured***
560 Ω Resistor		
1.0 kΩ Resistor		
220 Ω Resistor		
Equivalent Resistance		
The following are results for the circuit with the 220Ω resistor shorted.		
Equivalent Resistance		

Questions

1. Perform the error calculation for all entries in Table 7-2. Comment on whether the errors are acceptable.

2. How is total (equivalent) resistance in series circuits determined?

3. What effect does adding additional loads (resistors) in series circuit have on the total resistance?

4. What effect does adding loads (resistors) in series circuit have on the current in the circuit?

5. How will an open load (resistor) in a series circuit affect the circuit?

Bonus:

6. Given many of 2.0kΩ resistors, 2.2kΩ resistors, and 3.3kΩ resistors, construct a circuit that will have an equivalent resistance of 9.7kΩ.

Troubleshooting

Exercise #1

Consider a circuit diagram of a typical series connected Christmas lights (Figure 7-6).

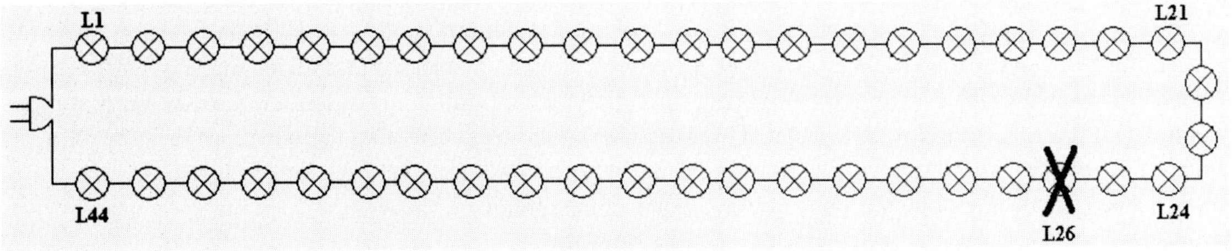

Figure 7-6

Light L26 is burned out and its filament is an open circuit. You, of course, do not know which light bulb is burned out, you just know that the lights do not light up at all. One way of fixing the problem is to replace on light bulb at a time until all the lights light up. This however, would take a long time. After performing this experiment and learning about resistance in series concepts, you can do better than that. Being equipped with the ohmmeter, suggest a course of actions that would quickly determine which light bulb is burned out. This method should work effectively, even though there may be more than one light bulb burned out.

8

Current in Series

Problem Statement

A behavior of current flow in a series circuit is what causes a typical strand of Christmas lights to completely turn off when at least one of the light bulbs burns out. By understanding this characteristic, you will be able to effectively design and analyze series circuits.

Concept

In a series circuit there is only one path in which current can flow. There are no branches through which current can leave, and no additional current can enter the series loop. With this in mind, whatever current leaves a source must also re-enter the source. Since no additional current can enter or leave the circuit, the current is the same everywhere in a series circuit. This means the current is the same through every load in series.

If you know the current anywhere in a series loop, you know the current everywhere in the series loop. For this reason, when analyzing a series circuit, it is a good idea to determine the current; you will then know something about every load in the series loop. Remember, current is measured with an ammeter. The ammeter is placed in series with the component through which current is to be measured. Also, keep in mind; an ammeter is a short to the circuit. This keeps the meter from affecting the operation of the circuit. In order to measure current in a series loop or circuit, it doesn't matter where the meter is placed in the loop. The current will still be the same at every point in the circuit.

Current may be predicted using Ohm's law. Once the voltage is known across any load, the current throughout the circuit can be determined. Current may also be determined using the source voltage, total resistance, and Ohm's law:

$$\mathbf{I=V/R}$$

Learning Objectives

Upon completion of this lab, the student will be able to:

1. Configure a circuit using eTCB schematic drawing and eTCB board.
2. Given voltage and resistance, predict the current in a series circuit.
3. Measure the current in multiple places in a series circuit.

Pre-Lab Procedure

1. Given the voltage V1=5.0V, predict current ***I*** in the series circuit shown in Figure 8-1. Record your result in Table 8-1.

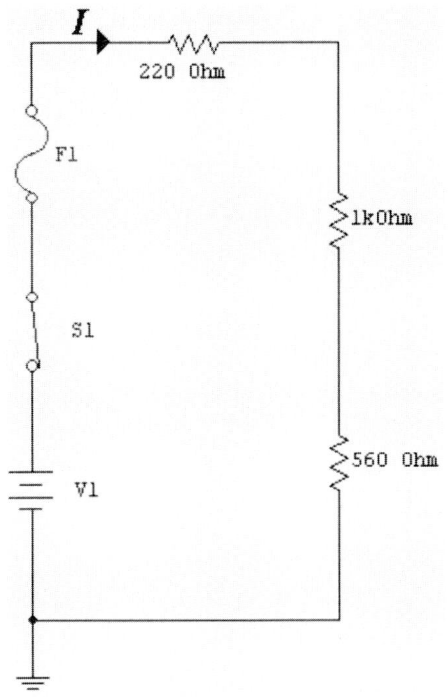

Figure 8-1

2. Repeat Pre-Lab Procedure 1 with the voltage V1=11.0V. Record your result in Table 8-1.

Lab Procedure

Before proceeding further, make sure that your eTCB is working correctly. Using eTCB Diagnostic VI, perform a self-diagnostic of eTCB circuitry.

Turn the switch S1 and S2 OFF in preparation for the lab.

Setup

Step #1 Using the eTCB Configuration VI, (DC module), first reset all relays and then configure the circuit shown in Figure 8-2.

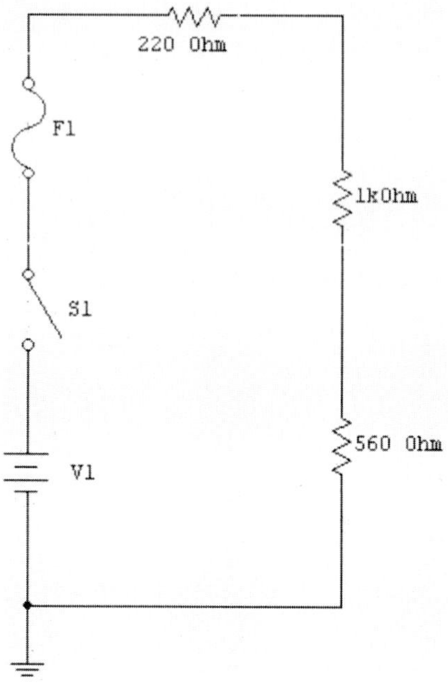

Figure 8-2

You will be asked to measure a DC current in this experiment, and as part of the DMM setup you need to "Null" the ammeter. This will minimize the error in DC current measurement. The "Nulling" procedure is very simple and needs to be performed only once per experiment. The procedure to "Null" the ammeter is as follows:

- Remove the test leads from the DMM terminals labeled "CURRENT".
- Open the Digital Multimeter VI and select DC Current measurement function.
- Click on the "Null" button.

Step #2 Using the Variable Power Supplies VI, set voltage source V1 to ***5.0V.*** Keep in mind that this setting is just a good indication of what the voltage of V1 is set to. Variable power supplies often need to be adjusted (tuned) in order to set their output voltage precisely. Keep the Variable Power Supplies VI ***open***; if you close it, all the variable voltage supplies will reset to 0V.

Step #3 Physically turn switch S1 on the eTCB board "ON". This will complete your series circuit. To verify that the switch is ON (closed) and that the fuse is not blown, use the Digital Multimeter VI to measure voltage at TP1 with respect to ground. Adjust the output of the variable power supplies, if necessary.

Discovery

Step #4 Using the Digital Multimeter VI, measure the current through the **560 Ω** resistor. Record this reading in Table 8-1 (column – Measured). To perform this measurement, you need to insert the ammeter in series with the **560 Ω** resistor. Removing the jumper **J7** creates a break in the circuit where the ammeter can be inserted. Refer to Figure 8-3 for ammeter placement information. After the measurement is performed, make sure that every jumper is placed back in its place.

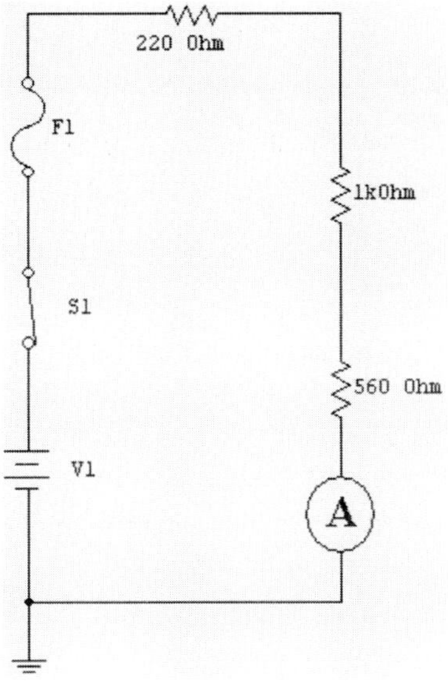

Figure 8-3

Step #5 Using the Digital Multimeter VI, measure the current through the **1.0 kΩ** resistor. Record this reading in Table 8-1 (column – Measured). To perform this measurement, you need to insert the ammeter in series with the **1.0 kΩ** resistor. Removing the jumper **J6** creates a break in the circuit where the ammeter can be inserted. Refer to Figure 8-4 for ammeter placement information. After the measurement is performed, make sure that every jumper is placed back in its place.

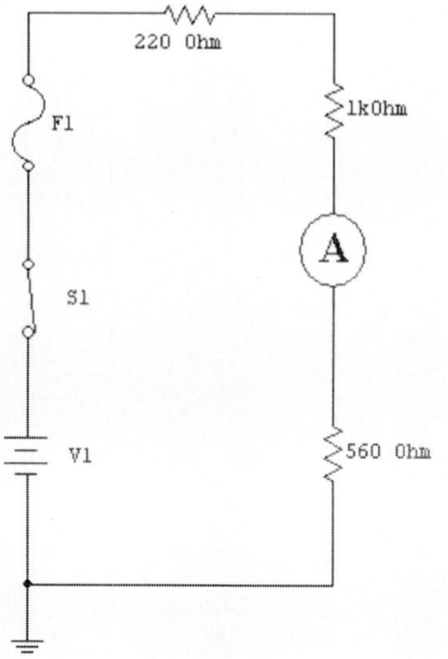

Figure 8-4

Step #6 Using the Digital Multimeter VI, measure the current through the **220 Ω** resistor. Record this reading in Table 8-1 (column – Measured). To perform this measurement, you need to insert the ammeter in series with the **220 Ω** resistor. Removing the jumper **I2** creates a break in the circuit where the ammeter can be inserted. Refer to Figure 8-5 for ammeter placement information. After the measurement is performed, make sure that every jumper is placed back in its place.

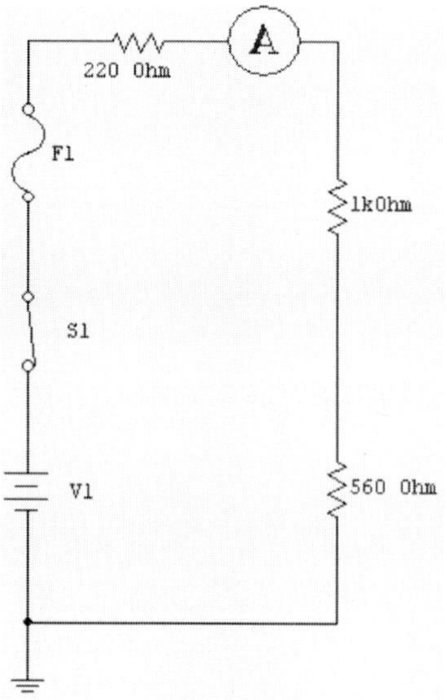

Figure 8-5

Note: We are now going to examine the effects of different supply voltage on the current in the series circuit. We are going to use the same circuit diagram, but rather than setting V1=5.0V, we are going to set it at ***V1=11.0V.***

Step #7 Using the Variable Power Supplies VI, set voltage source V1 to ***11.0V.*** Using the Digital Multimeter VI, measure voltage at TP1 with respect to ground. Adjust the output of the variable power supplies, if necessary.

Step #8 Using the Digital Multimeter VI, measure the current through the **560 Ω** resistor. Record this reading in Table 8-1 (column – Measured). Removing the jumper **J7** creates a break in the circuit where the ammeter can be inserted. Refer to Figure 8-3 for ammeter placement information. After the measurement is performed, make sure that every jumper is placed back in its place.

Step #9 Using the Digital Multimeter VI, measure the current through the **1.0 kΩ** resistor. Record this reading in Table 8-1 (column – Measured). Removing the jumper **J6** creates a break in the circuit where the ammeter can be inserted. Refer to Figure 8-4 for ammeter placement information. After the measurement is performed, make sure that every jumper is placed back in its place.

Step #10 Using the Digital Multimeter VI, measure the current through the **220 Ω** resistor. Record this reading in Table 8-1 (column – Measured). Removing the jumper **I2** creates a break in the circuit where the ammeter can be inserted. Refer to Figure 8-5 for ammeter placement information. After the measurement is performed, make sure that every jumper is placed back in its place.

This concludes the procedure for this experiment.

Table 8-1

Resistance	*Predicted Current*	*Measured Current*
V1=5.0V		
560 Ω		
1.0 kΩ		
220 Ω		
V1=11.0V		
560 Ω		
1.0 kΩ		
220 Ω		

Questions

1. Perform the error calculation for all entries in Table 8-1. Comment on whether the errors are acceptable.

2. How are the currents through each resistive load in a series circuit related to the resistance of each load?

3. In a series circuit, what is the relationship of the total current supplied by the power source to the currents through each resistive load?

4. How would increasing the total (equivalent) resistance in the circuit affect the total current supplied by the power source?

Bonus:

5. How many current paths exist in a series circuit?

9

Voltage in Series

Problem Statement

Not infrequently a situation exiswts where a certain voltage value is needed which happens to be below the voltage levels available in a particular design. An example of that scenario may be one where an expansive and very accurate reference voltage device is providing a single accurate reference voltage. However, a particular design also needs an accurate reference voltage, which happens to be of lesser voltage value than the one sourced by the available device. One solution is to simply add an additional reference voltage device (Circuit #1; refer to Figure 9-1). This solution is often cost prohibitive. The alternative solution is to set up an accurate voltage divider circuit (Circuit #2; refer to Figure 9-1). This circuit will divide the available reference voltage, and therefore establish the additional desired reference voltage. In this experiment, you will gain experience working with voltage dividers.

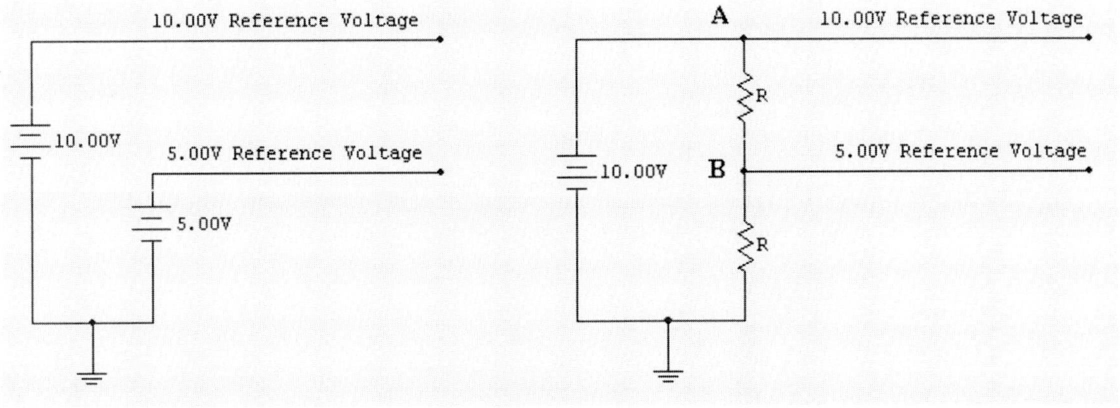

Circuit #1 Circuit #2

Figure 9-1

Note of caution: This solution is effective when there is relatively very little current drawn away from the voltage divider circuit at points A and B.

Concept

The voltage applied to a series circuit is applied across the entire circuit. It is not applied across each load. It is distributed across each load such that the sum of all the voltages across each load will be equal the applied voltage. This is explained by Kirchhoff's voltage law: *the sum of all voltage drops in series equals the applied voltage,* or more generally, *the sum of all potential differences in series circuits including the source equals zero.* This may also be stated in equation form:

$$V_{R1} + V_{R2} + \dots V_{RN} = V_{Source} \quad or \quad V_{R1} + V_{R2} + \dots V_{RN} - V_{Source} = 0$$

The voltage developed across a load in series depends upon the amount of current in the circuit and the opposition to the flow of current in that load. The greater the resistance to current flow the more force is necessary to cause the current to flow through the load, therefore the greater the voltage developed across the load.

The voltage developed across a load in series circuit may be determined in two ways, as follows.

1. *With the use of Ohm's law equation:*

$$V=IR$$

This further demonstrates the direct relationship between voltage across the load, current through the load and load resistance. From this equation, you can see that if the resistance is increased, more voltage is required to cause the same current to flow through the load.

2. *With the use of voltage divider equation:*

$$(R_N/R_{Total})V_{Source} = V_{RN}$$

From this equation, it can be demonstrated that if the fractional part of the total resistance a single load represents, it will require the same fractional part of the applied voltage to cause the current to flow through the load.

Learning Objectives

Upon completion of this lab, the student will be able to:

1. Predict and measure voltages across all components in a single voltage source series circuit.
2. Predict individual load voltage drops using the voltage divider equation in a single voltage source series circuit.

Pre-Lab Procedure

1. Using Ohm's law equation, predict voltage drops across each load resistor in the circuit in Figure 9-2. Let the voltage V1=10.0V. Record your results in Table 9-1 (column – Predicted Voltage).

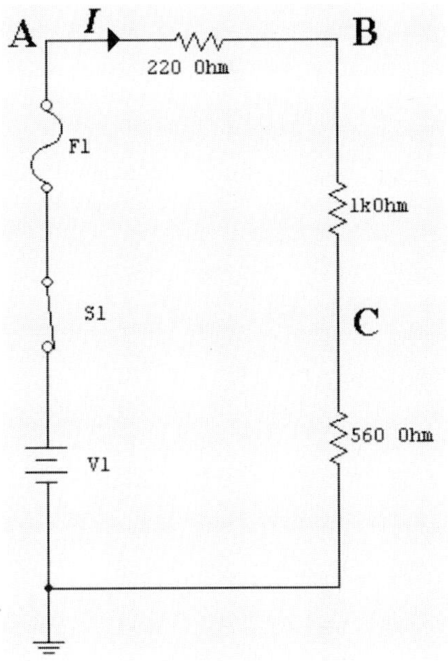

Figure 9-2

2. Using voltage divider equation, predict voltage drops across each load resistor in the circuit in Figure 9-2. Record your results in Table 9-2 (column – Predicted Voltage).

3. Using above results, predict voltage at point A, B, and C with respect to ground in the circuit in Figure 9-2. Record your results in Table 9-3 (column – Predicted Voltage).

Lab Procedure

Before proceeding further, make sure that your eTCB is working correctly. Using eTCB Diagnostic VI, perform a self-diagnostic of eTCB circuitry.

Turn the switch S1 and S2 OFF in preparation for the lab.

Setup

Step #1 Using the eTCB Configuration VI, (DC Module), first reset all relays and then configure the circuit in Figure 9-3.

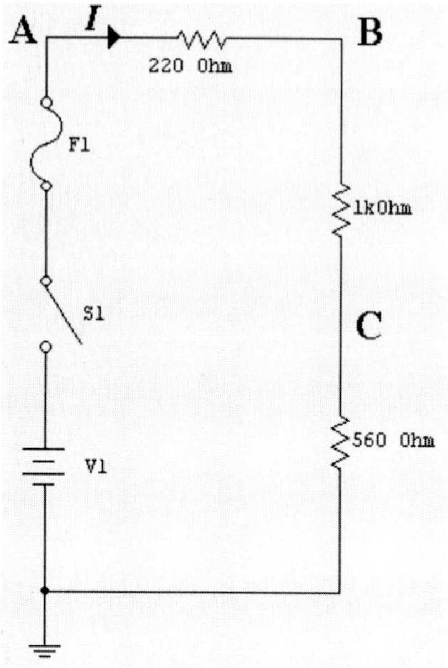

Figure 9-3

Step #2 Using the Variable Power Supplies VI, set voltage source V1 to ***10.0V.*** Keep in mind that this setting is just a good indication of what the voltage of V1 is set to. Variable power wupplies often need to be adjusted (tuned) in order to set their output voltage precisely. Keep the Variable Power Supplies VI ***open***; if you close it, all the variable voltage supplies will reset to 0V.

Step #3 Physically turn switch S1 on the eTCB board "ON". This will complete your series circuit. To verify that the switch is ON (closed) and that the fuse is not blown, use the Digital Multimeter VI to measure voltage at TP1 with respect to ground. Adjust the output of the variable power supplies, if necessary.

Discovery

Step #4 Using the Digital Multimeter VI, measure voltage across the 560 Ω resistor. Record this reading in Table 1-1 and Table 9-2 (column – Measured Voltage).

Step #5 Using the Digital Multimeter VI, measure voltage across the 1.0 kΩ resistor. Record this reading in Table 9-1 and Table 9-2 (column – Measured Voltage).

Step #6 Using the Digital Multimeter VI, measure voltage across the 220 Ω resistor. Record this reading in Table 9-1 and Table 9-2 (column – Measured Voltage).

Step #7 Using the Digital Multimeter VI, measure voltage at Point A with respect to ground Record this reading in Table 9-3 (column – Measured Voltage).

Step #8 Using the Digital Multimeter VI, measure voltage at Point B with respect to ground Record this reading in Table 9-3 (column – Measured Voltage).

Step #9 Using the Digital Multimeter VI, measure voltage at Point C with respect to ground Record this reading in Table 9-3 (column – Measured Voltage).

This concludes the procedure for this experiment.

Table 9-1

Ohm's Law Predictions

Load Resistor	***Predicted Voltage***	***Measured Voltage***
560 Ω		
1.0k Ω		
220 Ω		

Table 9-2

Voltage Divider Predictions

Load Resistor	***Predicted Voltage***	***Measured Voltage***
560 Ω		
1.0k Ω		
220 Ω		

Table 9-3

Kirchhoff's Voltage Law

	Predicted Voltage	***Measured Voltage***
@ Point A		
@ Point B		
@ Point C		

Questions

1. Perform the error calculation for all entries in Table 9-1, 9-2, and 9-3. Comment on whether the errors are acceptable.

2. What is the relationship between resistor value and voltages across it. Use measured voltage values from Table 9-1 to support your answer.

3. Considering resistor value, which resistor in a series circuit will have the greatest voltage drop? Why? Use measured voltage value from Table 9-2 to support your answer.

4. Considering resistor value , which resistor in a series circuit will have the smallest voltage drop? Why? Use measured voltage from Table 9-2 to support your answer.

5. What are the advantages and disadvantages of using Ohm's law equation vs. voltage divider equation to determine voltage across an individual resistor?

6. Use measured resistor voltage values from Table 9-2 to verify Kirchhoff's voltage equations in two different ways.

Bonus:

7. Use measured voltage at Point B from Table 9-3 and measured resistor voltage from Table 9-2 to verify Kirchhoff's voltage law.

10

Power in Series

Problem Statement

Consider Christmas lights where each light bulb is connected in series. Each light bulb is rated at about 7 W and it offers a resistance of about 15_. The challenge is to build Christmas lights where no one light bulb is overrated in terms of power handling capacity. Understanding how power is distributed in a series circuit will help in building Christmas lights with the correct number of light bulbs.

Learning Objectives

Upon completion of this lab, the student will be able to:

1. Configure a circuit using eTCB schematic drawing and eTCB board.
2. Given any two electrical parameters voltage, current, or resistance, determine the power dissipation of an electrical load in series circuit.
3. Given any two electrical parameters voltage, current, or resistance, determine the total power dissipation in series circuit.

Pre-Lab Procedure

1. Use all three different power equations to predict the power dissipated in **each resistor** (Figure 10-1). Perform this calculation for supply voltage V1=7.0V. Record your results in Table 10-1.

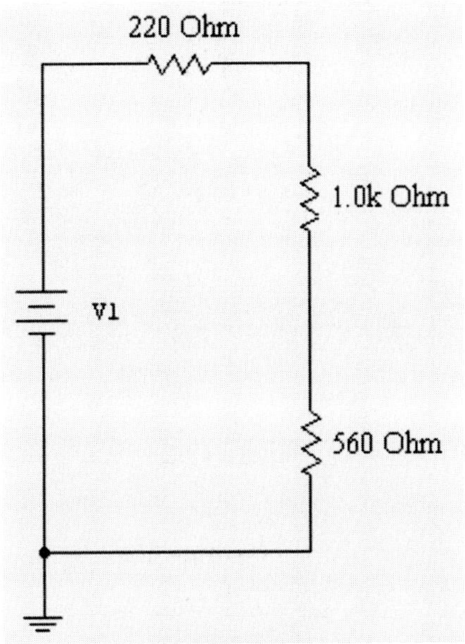

Figure 10-1

2. Use all three different power equations to predict the **total power** dissipated in the series circuit (Figure 10-1). Perform this calculation for supply voltage V1=7.0V. Record your results in Table 10-1.

3. Repeat Pre-Lab Procedure #1 and #2 for supply voltage V1=11.0V. Record your results in Table 10-2.

Lab Procedure

Before proceeding further, make sure that your eTCB is working correctly. Using eTCB Diagnostic VI, perform a self-diagnostic of eTCB circuitry.

Turn the switch S1 and S2 OFF in preparation for the lab.

Setup

Step #1 Using the eTCB Configuration VI, (DC Module), first reset all relays and then configure the circuit in Figure 10-2 .

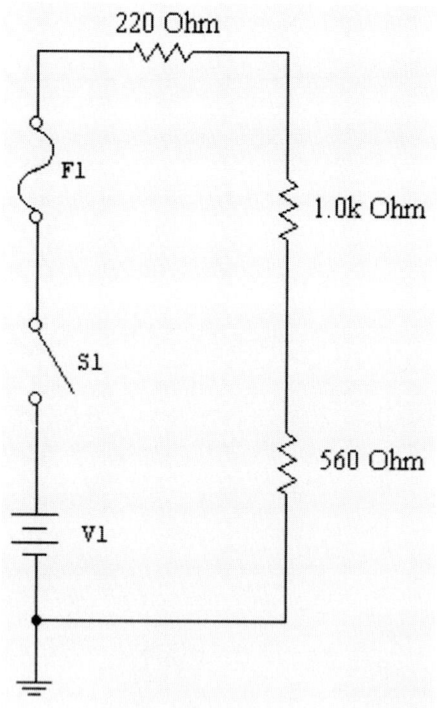

Figure 10-2

You will be asked to measure a DC current in this experiment, and as part of the DMM setup you need to "Null" the ammeter. This will minimize the error in DC current measurement. The "Nulling" procedure is very simple and needs to be performed only once per experiment. The procedure to "Null" the ammeter is as follows:

- Remove the test leads from the DMM terminals labeled "CURRENT".
- Open the Digital Multimeter VI and select DC Current measurement function.
- Click on the "Null" button.

Step #2 Using the Variable Power Supplies VI, set voltage source V1 to ***7.0V***. Keep in mind that this setting is just a good indication of what the voltage of V1 is set to. Variable power supplies often need to be adjusted (tuned) in order to set their output voltage precisely. Keep the Variable Power Supplies VI ***open***; if you close it, all the variable voltage supplies will reset to 0V.

Step #3 Physically turn switch S1 on the eTCB board "ON". This will complete your series circuit. To verify that the switch is ON (closed) and that the fuse is not blown, use the Digital Multimeter VI to measure voltage at TP1 with respect to ground. Adjust the output of the variable power supplies, if necessary. Record this reading in Table 10-1. Refer to Figure 10-3 for help in placing the voltmeter.

Discovery

Step #4 Using the Digital Multimeter VI, measure current in the circuit. Record this reading in Table 10-1. To perform this measurement, you need to insert the ammeter in series. Removing the jumper **J7** creates a break in the circuit where the ammeter can be inserted. Refer to Figure 10-3. After the measurement is performed, make sure that every jumper is placed back in its place.

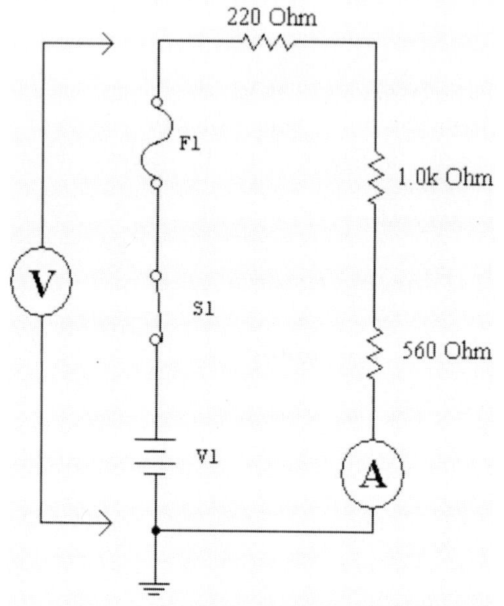

Figure 10-3

Carefully, touch each resistor in the circuit to get a sense of how warm it feels.

Step #5 Using the Digital Multimeter VI, measure voltage across the **560 Ω** resistor. Record this reading in Table 10-1.

Step #6 Using the Digital Multimeter VI, measure voltage across the **1.0k Ω** resistor. Record this reading in Table 10-1.

Step #7 Using the Digital Multimeter VI, measure voltage across the **220 Ω** resistor. Record this reading in Table 10-1.

Note: Now we are going to examine the effects of different supply voltage on the power dissipation. We are going to use the same circuit diagram, but rather than setting supply voltage V1 to 7.0V, we are going to set it to ***11.0V.***

Step #8 Using the Variable Power Supplies VI, set voltage source V1 to ***11.0V.*** Keep in mind that this setting is just a good indication of what the voltage of V1 is set to. Variable power supplies often need to be adjusted (tuned) in order to set their output voltage precisely. Keep the Variable Power Supplies VI ***open***; if you close it, all the variable voltage supplies will reset to 0V.

Using the Digital Multimeter VI, measure voltage at TP1 with respect to ground. Adjust the output of the variable power supplies, if necessary. Record this reading in Table 10-2.

Step #9 Using the Digital Multimeter VI, measure current in the circuit. Record this reading in Table 10-2. To perform this measurement, you need to insert the ammeter in series. Removing the jumper **J7** creates a break in the circuit where the ammeter can be inserted. Refer to Figure 10-3. After the measurement is performed, make sure that every jumper is placed back in its place.

Step #10 Using the Digital Multimeter VI, measure voltage across the **560 Ω** resistor. Record this reading in Table 10-2.

Step #11 Using the Digital Multimeter VI, measure voltage across the **1.0k Ω** resistor. Record this reading in Table 10-2.

Step #12 Using the Digital Multimeter VI, measure voltage across the **220 Ω** resistor. Record this reading in Table 10-2.

This concludes the procedure for this experiment.

Table 10-1

	Predicted	***Measured***
Supply Voltage (VI)	**7.0V**	
Current		
Total Power (VI)		*
Total Power (V^2/R)		*
Total Power (I^2R)		*
	560 Ω Resistor	
Resistor Voltage		
Resistor Power (VI)		*
Resistor Power (V^2/R)		*
Resistor Power (I^2R)		*
	1.0 kΩ Resistor	
Resistor Voltage		
Resistor Power (VI)		*
Resistor Power (V^2/R)		*
Resistor Power (I^2R)		*
	220 Ω Resistor	
Resistor Voltage		
Resistor Power (VI)		*
Resistor Power (V^2/R)		*
Resistor Power (I^2R)		*

	*** Calculated based on measured values**

Table 10-2

	Predicted	*Measured*
Supply Voltage (V1)	**11.0V**	
Current		
Total Power (VI)		*
Total Power (V^2/R)		*
Total Power (I^2R)		*
	560 Ω Resistor	
Resistor Voltage		
Resistor Power (VI)		*
Resistor Power (V^2/R)		*
Resistor Power (I^2R)		*
	1.0 kΩ Resistor	
Resistor Voltage		
Resistor Power (VI)		*
Resistor Power (V^2/R)		*
Resistor Power (I^2R)		*
	220 Ω Resistor	
Resistor Voltage		
Resistor Power (VI)		*
Resistor Power (V^2/R)		*
Resistor Power (I^2R)		*

	*** Calculated based on measured values**

Questions

1. Perform the error calculation for all entries in Table 10-1, and Table 10-2. Comment on whether the errors are acceptable.

2. What effect does doubling of voltage across a resistive load have on power dissipation?

3. What effect does doubling of resistance value of a restive load have on power dissipation?

4. What effect does doubling of current through a restive load have on power dissipation?

5. Since current is the same in a series circuit, what determines the power dissipation in each of the resistive loads?

6. How does the total power in a series circuit relate the individual powers dissipated in the resistive loads?

Bonus:

7. Given that the maximum supply voltage to the Christmas tree is 180V, and each light bulb is rated at 7W and offers 10_ of resistance, calculate the minimum number of light bulbs that have to be used. (Refer to Figure 10-4.)

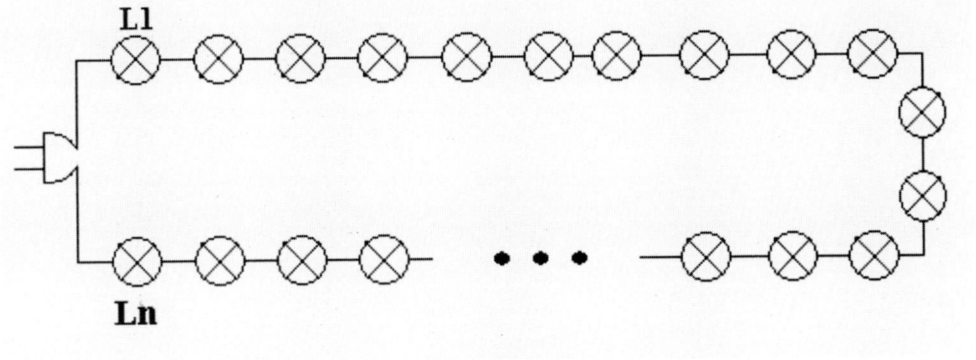

Figure 10-4

Troubleshooting

Exercise #1

Determine which of the following resistors may be damaged by overheating. Justify your answer with power calculation.

a) 560 Ω resistor rated at 1/2W with 25V across it.
b) 820 Ω resistor rated at 1/2W with 12V across it.
c) 220 Ω resistor rated at 1/4W with 40mA through it.
d) 470 Ω resistor rated at 1/4W with 20mA through it.
e) Resistor rated at 1/2W with 50mA through it and 12V across it.
f) Resistor rated at 1/4W with 10mA through it and 12V across it.

Exercise #2

Resistors Figure 10-5 are all rated at 1/4W; they have been connected to voltage supply V1=50V. The current I in the circuit is zero. What caused the current to be zero?

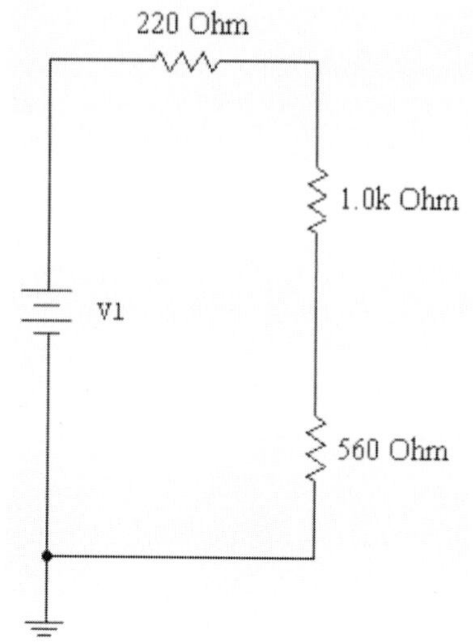

Figure 10-5

11

Resistance in Parallel

Problem Statement

At times a resistance-sensitive circuit may call for a resistance value that is not a standard value. Understanding how equivalent resistance is determined in parallel circuits will help in building a circuit with multiple standard value resistors, which together will offer the desired equivalent resistance.

Concept

A parallel circuit is any circuit with more than one path for current. Each parallel branch is a unique path for current to flow. The more current paths there are, the greater the current and the lower the total resistance of the circuit. Because there is an inverse relationship between current and resistance; the greater the current, the lower the resistance. Note that the total resistance in parallel will always be less than the smallest parallel branch resistance.

There are a number of ways to determine the resistance in a parallel circuit. When current is known, resistance may be determined using Ohm's law, $R = V \div I$. Resistance may also be determined by adding all of the conductance in parallel and converting to resistance. Conductance (G) is the reciprocal of resistance:

$$G=1/R$$

Total conductance in parallel is the sum of all parallel branch conductance. Total conductance is reciprocated to convert to total resistance:

$$Rt=1/Gt$$

Applying this concept to circuit in Figure 11.1, we can state that:

$$Geq=G1+G2+G3$$

Or, converted to resistance:

$$1/Req=1/R1+1/R2+1/R3$$

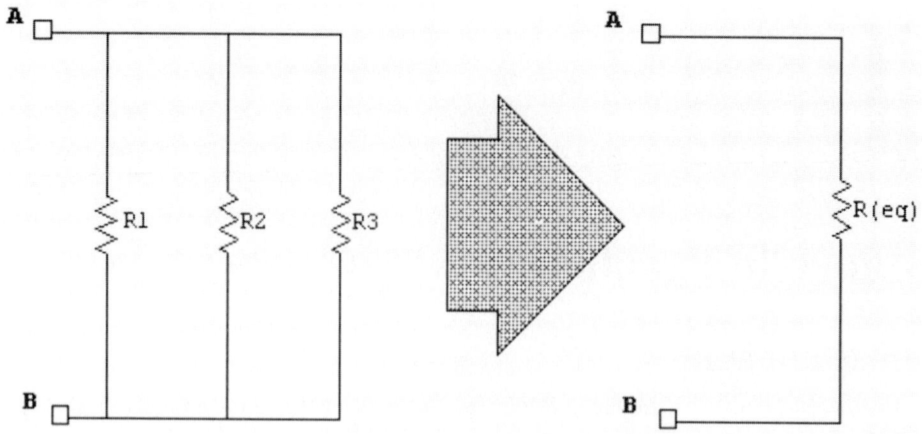

Figure 11-1

Learning Objectives

Upon completion of this lab, the student will be able to:

1. Configure a circuit using eTCB schematic drawing and eTCB board.
2. Predict and measure the resistance in a parallel circuit.

Pre-Lab Procedure

1. Given the circuit in Figure 11-2 and resistors color code in Table 11-1, determine the resistance of each of the resistors as well as the equivalent resistance seen between points A and B. Record the equivalent resistance result in Table 11-2.

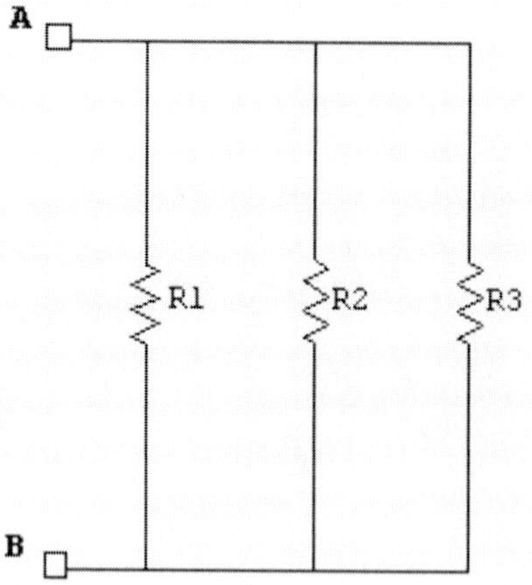

Figure 11-2

Table 11-1

	First Band	*Second Band*	*Third Band*	*Forth Band*	*Rated Resistance*	*Tolerance*
R1	*Yellow*	*Violet*	*Brown*	*Gold*		
R2	*Green*	*Blue*	*Brown*	*Gold*		
R3	*Brown*	*Red*	*Red*	*Gold*		

2. Repeat Pre-Lab Procedure 1 with resistor R1 being opened.

Lab Procedure

Before proceeding further, make sure that your eTCB is working correctly. Using eTCB Diagnostic VI, perform a self-diagnostic of eTCB circuitry.

Turn the switch S1 and S2 OFF in preparation for the lab.

Setup

Step #1 Using the eTCB Configuration VI, (DC Module), first reset all relays and then configure the circuit in Figure 11-3.

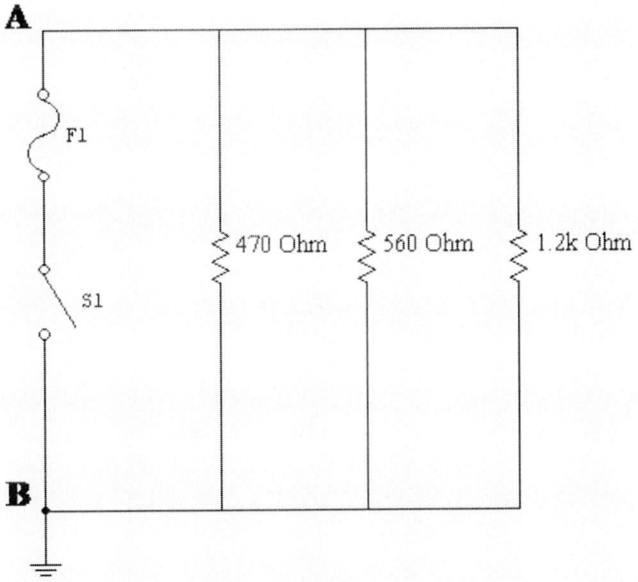

Figure 11-3

Note: Make sure that switch S1 remains "open" "OFF" and all the power supplies are completely isolated from any components on the eTCB board.

Discovery

Step #2 Using the Digital Multimeter VI, measure the resistance of **1.2k Ω** resistor. To isolate this resistor from the rest of the circuit, remove jumper **J4.** Record this reading in Table 11-2 (column – Measured). After the measurement is performed, place the jumper back in its place.

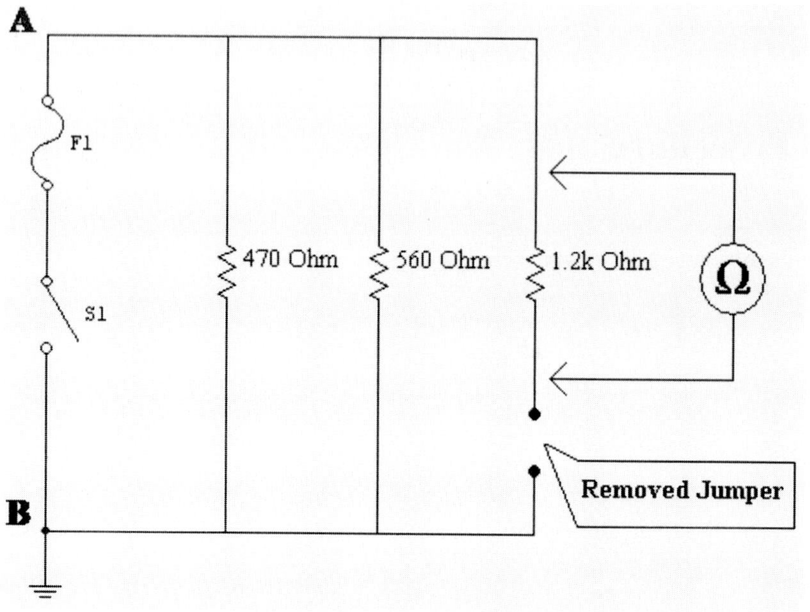

Figure 11-4

Step #3 Using the Digital Multimeter VI, measure the resistance of **560 Ω** resistor. To isolate this resistor from the rest of the circuit, remove jumper **J3.** Record this reading in Table 11-2 (column – Measured). After the measurement is performed, place the jumper back in its place.

Step #4 Using the Digital Multimeter VI, measure the resistance of **470 Ω** resistor. To isolate this resistor from the rest of the circuit, remove jumper **J2**. Record this reading in Table 11-2 (column – Measured). After the measurement is performed, place the jumper back in its place.

Step #5 Using the Digital Multimeter VI, measure the equivalent resistance of the parallel circuit seen from points **A and B.** Record this reading in Table 11-2 (column – Measured).

Step #6 Using the eTCB Configuration VI, (DC Module), configure the circuit in Figure 11-5. This circuit has 470 Ω resistor which is opened.

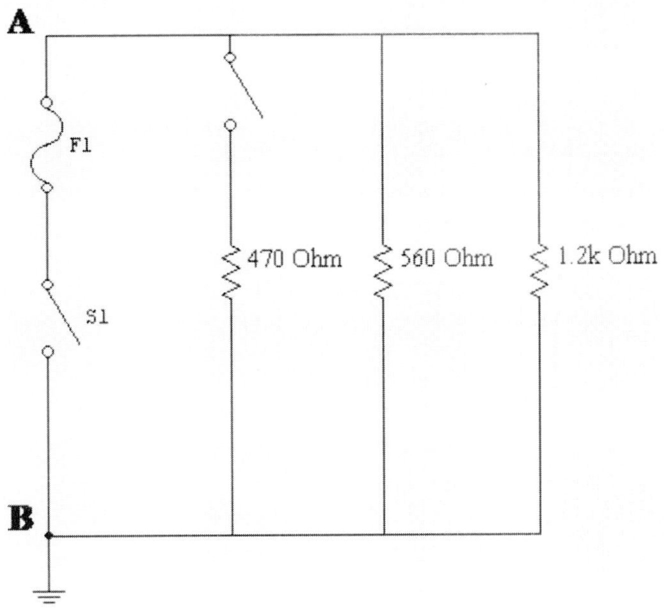

Figure 11-5

Note: Make sure that switch S1 remains "open" "OFF" and all the power supplies are completely isolated from any components on the eTCB board.

Step #7 Using the Digital Multimeter VI, measure the equivalent resistance of the parallel circuit seen from points **A and B.** Record this reading in Table 11-2 (column – Measured).

This concludes the procedure for this experiment.

Table 11-2

	Predicted	***Measured***
1.2k Ω Resistor		
560 Ω Resistor		
470 Ω Resistor		
Equivalent Resistance		
The following are results for the circuit with the 470 Ω resistor opened.		
Equivalent Resistance		

Questions

1. Perform the error calculation for all entries in Table 11-2. Comment on whether the errors are acceptable.

2. How is total (equivalent) resistance in parallel circuits determined?

3. What effect does adding additional loads (resistors) in parallel circuit have on the total resistance?

4. How will an open load (resistor) in a parallel circuit affect the circuit?

5. How will a shorted load (resistor) in a parallel circuit affect the circuit?

Bonus:

1. Given many of 2.0kΩ resistors, and 3.3kΩ resistors, construct a circuit that will have an equivalent resistance of 904.1 Ω.

12

Voltage in Parallel

Concept

The voltage across parallel branches is the same for all parallel branches. The current in each parallel branch is independent of every other parallel branch current. As additional branches are added, the voltage across all branches remains the same. For this reason, as branches are added or removed, the operation of the remaining branches remains unchanged.

Learning Objectives

Upon completion of this lab, the student will be able to:

1. Configure a circuit using eTCB schematic drawing and eTCB board.
2. Given supply voltage value, predict voltage drops on each resistor in a parallel circuit.
3. Measure the voltage in a parallel circuit.

Pre-Lab Procedure

1. Given the voltage V1=5.0V, predict the voltage across each resistor load in the parallel circuit shown in Figure 12-1. Record your result in Table 12-1.

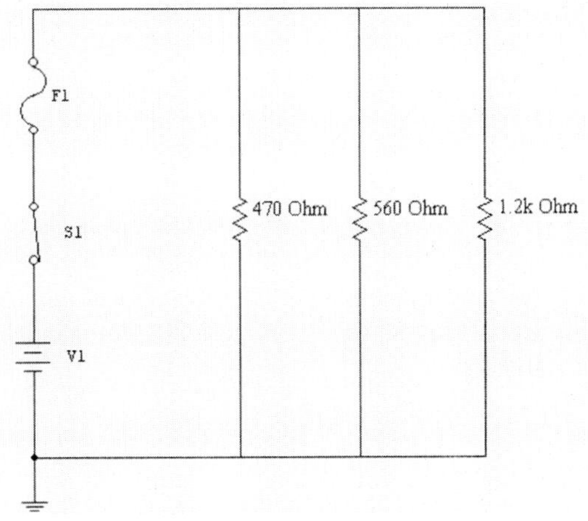

Figure 12-1

2. Repeat Pre-Lab Procedure 1 with the voltage V1=11.0V. Record your result in Table 12-1.

Lab Procedure

Before proceeding further, make sure that your eTCB is working correctly. Using eTCB Diagnostic VI, perform a self-diagnostic of eTCB circuitry.

Turn the switch S1 and S2 OFF in preparation for the lab.

Setup

Step #1 Using the eTCB Configuration VI, (DC module), first reset all relays and then configure the circuit shown in Figure 12-2.

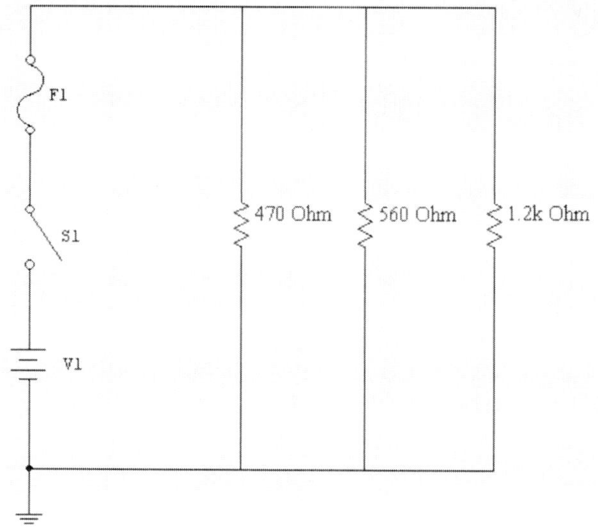

Figure 12-2

Step #2 Using the Variable Power Supplies VI, set voltage source V1 to ***5.0V***. Keep in mind that this setting is just a good indication of what the voltage of V1 is set to. Variable power supplies often need to be adjusted (tuned) in order to set their output voltage precisely. Keep the Variable Power Supplies VI ***open***; if you close it, all the variable voltage supplies will reset to 0V.

Step #3 Physically turn switch S1 on the eTCB board "ON". This will complete your parallel circuit. To verify that the switch is "ON" (closed) and that the fuse is not blown, use the Digital Multimeter VI to measure voltage at TP1 with respect to ground. Adjust the output of the variable power supplies, if necessary.

Discovery

Step #4 Using the Digital Multimeter VI, measure the voltage across the **1.2k Ω** resistor. Record this reading in Table 12-1 (column – Measured). Refer to Figure 12-3 for voltmeter placement information.

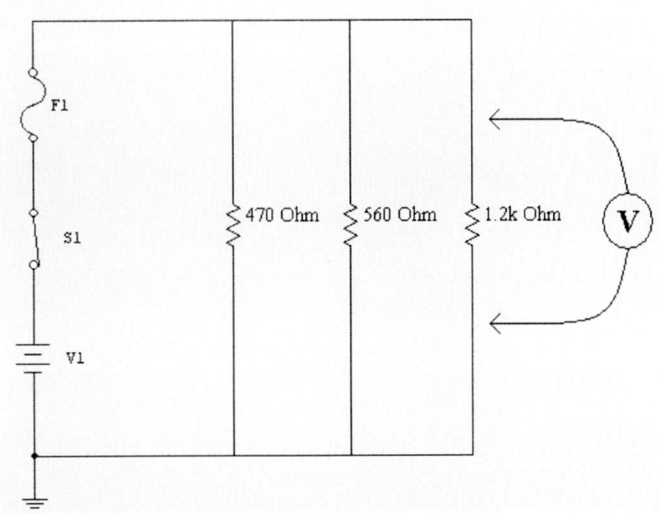

Figure 12-3

Step #5 Using the Digital Multimeter VI, measure the voltage across the **560 Ω** resistor. Record this reading in Table 12-1 (column – Measured).

Step #6 Using the Digital Multimeter VI, measure the voltage across the **470 Ω** resistor. Record this reading in Table 12-1 (column – Measured).

*Note: We are now going to examine the effects of different supply voltage on the voltage drops across the individual resistors in the parallel circuit. We are going to use the same circuit diagram, but rather than setting V1=5.0V, we are going to set it at **V1=11.0V.***

Step #7 Using the Variable Power Supplies VI, set voltage source V1 to ***11.0V.***
Using the Digital Multimeter VI, measure voltage at **TP1** with respect to ground. Adjust the output of the variable power supplies, if necessary.

Step #8 Using the Digital Multimeter VI, measure the voltage across the **1.2k Ω** resistor. Record this reading in Table 12-1 (column – Measured).

Step #9 Using the Digital Multimeter VI, measure the voltage across the **560 Ω** resistor. Record this reading in Table 12-1 (column – Measured).

Step #10 Using the Digital Multimeter VI, measure the voltage across the **470 Ω** resistor. Record this reading in Table 12-1 (column – Measured).

This concludes the procedure for this experiment.

Table 12-1

	Predicted Voltage	Measured Voltage
V1=5.0V		
1.2k Ω		
560 Ω		
470 Ω		
V1=11.0V		
1.2k Ω		
560 Ω		
470 Ω		

Questions

1. Perform the error calculation for all entries in Table 12-1. Comment on whether the errors are acceptable.

2. What general statement can be made about voltage across parallel branches?

3. What does adding additional loads in parallel do to the total voltage?

4. What does adding additional branches in parallel do to the current in original branches?

13

Current in Parallel

Concept

Total current in a parallel circuit is equal to the sum of all parallel branch currents. Each branch current may be determined using Ohm's law. Because the voltage across each parallel branch is the same, the branch currents are inversely related to the resistance of the branch. As branches are added to parallel circuits, the total current increases. This increase in total circuit current causes a decrease in total circuit resistance. The inverse occurs as the number of branches is decreased and total circuit current decreases, increasing the total circuit resistance. Adding and removing parallel branches has no effect on the other branch currents. Only the total current and total resistance is affected.

The current through a parallel branch may also be determined by the current divider equation:

Irn=It*Rt/Rn

This is an accurate determination of branch current no matter how many branches are in parallel. It should be noted that this is simply an application of Ohm's law. Total resistance multiplied by total current is equal to the applied voltage. The applied voltage is then divided by the branch resistance to determine the branch current.

Learning Objectives

Upon completion of this lab, the student will be able to:

1. Configure a circuit using eTCB schematic drawing and eTCB board.
2. Given voltage and resistance, predict the current in a parallel circuit.
3. Measure the current in multiple places in a parallel circuit.

Pre-Lab Procedure

1. Given the voltage V1=5.0V, predict current ***I*** in the parallel circuit shown in Figure 13-1. Record your result in Table 13-1.

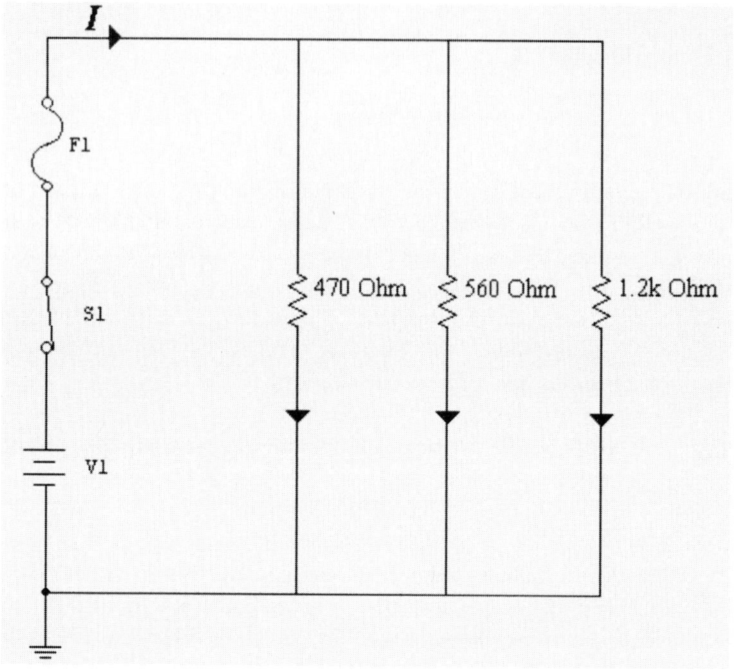

Figure 13-1

2. Repeat Pre-Lab Procedure 1 with the voltage V1=11.0V. Record your result in Table 13-1.

Lab Procedure

Before proceeding further, make sure that your eTCB is working correctly. Using eTCB Diagnostic VI, perform a self-diagnostic of eTCB circuitry.

Turn the switch S1 and S2 OFF in preparation for the lab.

Setup

Step #1 Using the eTCB Configuration VI, (DC Module), reset all relays and then configure the circuit in Figure 13-2.

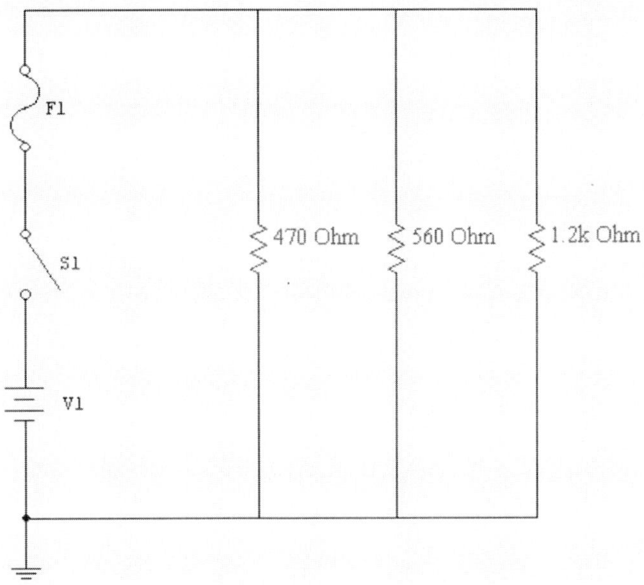

Figure 13-2

You will be asked to measure a DC current in this experiment, and as part of the DMM setup you need to "Null" the ammeter. This will minimize the error in DC current measurement. The "Nulling" procedure is very simple and needs to be performed only once per experiment. The procedure to "Null" the ammeter is as follows:

- Remove the test leads from the DMM terminals labeled "CURRENT".
- Open the Digital Multimeter VI and select DC Current measurement function.
- Click on the "Null" button.

Step #2 Using the Variable Power Supplies VI, set voltage source V1 to ***5.0V.*** Keep in mind that this setting is just a good indication of what the voltage of V1 is set to. Variable power supplies often need to be adjusted (tuned) in order to set their output voltage precisely. Keep the Variable Power Supplies VI ***open***; if you close it, all the variable voltage supplies will reset to 0V.

Step #3 Physically turn switch S1 on the eTCB board "ON". This will complete your parallel circuit. To verify that the switch is ON (closed) and that the fuse is not blown, use the Digital Multimeter VI to measure voltage at TP1 with respect to ground. Adjust the output of the variable power supplie,s if necessary.

Discovery

Step #4 Using the Digital Multimeter VI, measure the current through the **1.2k Ω** resistor. Record this reading in Table 13-1 (column – Measured). To perform this measurement, you need to insert the ammeter in series with the **1.2K Ω** resistor. Removing the jumper **J4** creates a break in the circuit where the ammeter can be inserted. Refer to Figure 13-3 for ammeter placement information. After the measurement is performed, make sure that every jumper is placed back in its place.

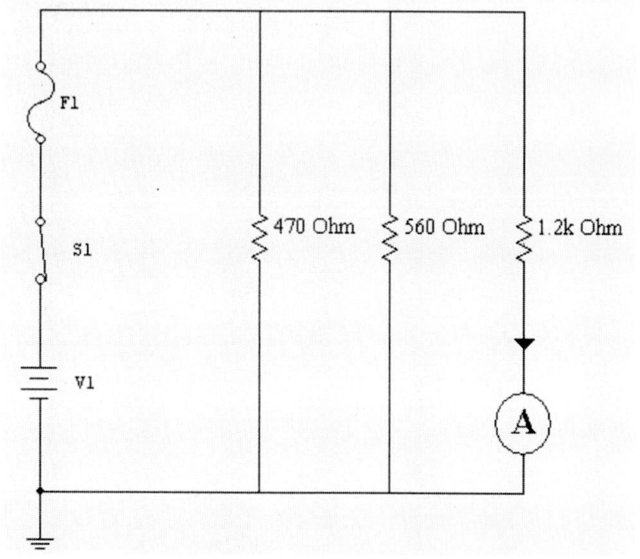

Figure 13-3

Step #5 Using the Digital Multimeter VI, measure the current through the **560 Ω** resistor. Record this reading in Table 13-1 (column – Measured). To perform this measurement, you need to insert the ammeter in series with the **560 Ω** resistor. Removing the jumper **J3** creates a break in the circuit where the ammeter can be inserted. Refer to Figure 13-4 for ammeter placement information. After the measurement is performed, make sure that every jumper is placed back in its place.

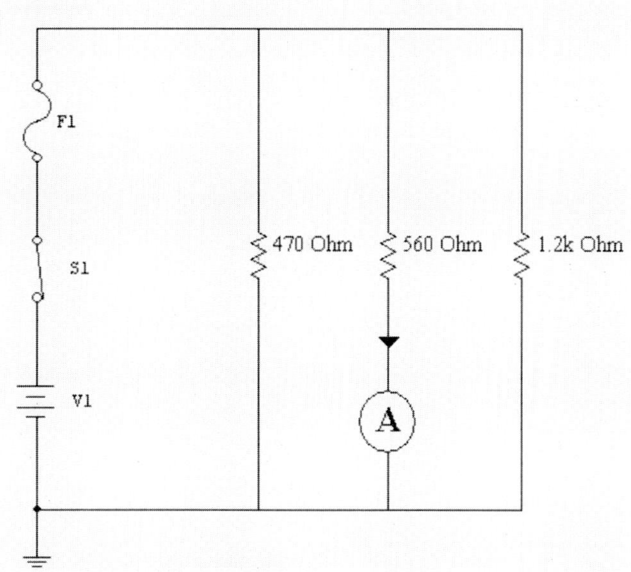

Figure 13-4

Step #6 Using the Digital Multimeter VI, measure the current through the **470 Ω** resistor. Record this reading in Table 13-1 (column – Measured). To perform this measurement, you need to insert the ammeter in series with the **470 Ω** resistor. Removing the jumper **J2** creates a break in the circuit where the ammeter can be inserted. Refer to Figure 13-5 for ammeter placement information. After the measurement is performed, make sure that every jumper is placed back in its place.

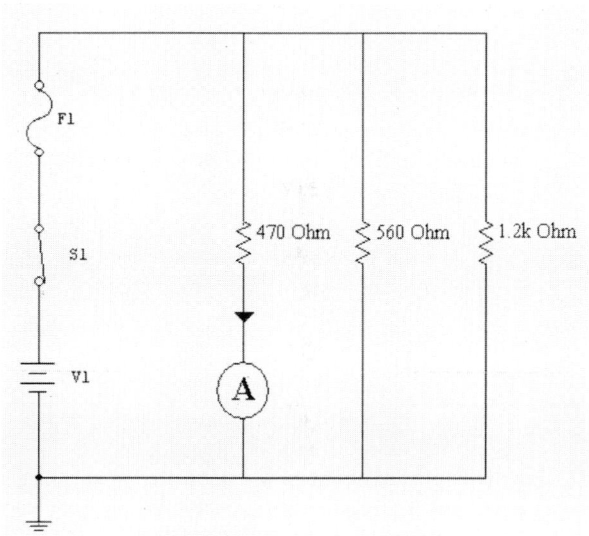

Figure 13-5

Step #7 Using the Digital Multimeter VI, measure the total current into the parallel circuit. Record this reading in Table 13-1 (column – Measured). Removing the jumper **I1** creates a break in the circuit where the ammeter can be inserted. Refer to Figure 13-6 for ammeter placement information. After the measurement is performed, make sure that every jumper is placed back in its place.

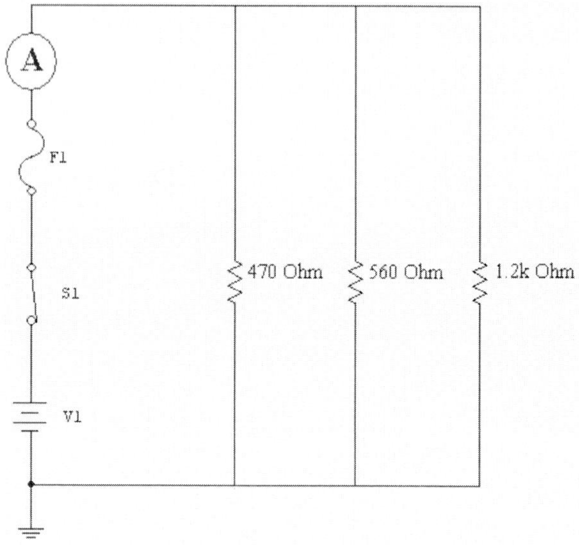

Figure 13-6

*Note: We are now going to examine the effects of different supply voltage on the currents in the parallel circuit. We are going to use the same circuit diagram, but rather than setting V1=5.0V, we are going to set it at **V1=11.0V.***

Step #8 Using the Variable Power Supplies VI, set voltage source V1 to ***11.0V.*** Using the Digital Multimeter VI, measure voltage at TP1 with respect to ground. Adjust the output of the variable power supplies, if necessary.

Step #9 Using the Digital Multimeter VI, measure the current through the **1.2k Ω** resistor. Record this reading in Table 13-1 (column – Measured). To perform this measurement, you need to insert the ammeter in series with the **1.2K Ω** resistor. Removing the jumper **J4** creates a break in the circuit where the ammeter can be inserted. Refer to Figure 13-3 for ammeter placement information. After the measurement is performed, make sure that every jumper is placed back in its place.

1. Step #10 Using the Digital Multimeter VI, measure the current through the **560 Ω** resistor. Record this reading in Table 13-1 (column – Measured). To perform this measurement, you need to insert the ammeter in series with the **560 Ω** resistor. Removing the jumper **J3** creates a break in the circuit where the ammeter can be inserted. Refer to Figure 13-4 for ammeter placement information. After the measurement is performed, make sure that every jumper is placed back in its place.

   Step #11 Using the Digital Multimeter VI, measure the current through the **470 Ω** resistor. Record this reading in Table 13-1 (column – Measured). To perform this measurement. you need to insert the ammeter in series with the **470 Ω** resistor. Removing the jumper **J2** creates a break in the circuit where the ammeter can be inserted. Refer to Figure 13-5 for ammeter placement information. After the measurement is performed, make sure that every jumper is placed back in its place.

Step #12 Using the Digital Multimeter VI, measure the total current into the parallel circuit. Record this reading in Table 13-1 (column – Measured). Removing the jumper **I1** creates a break in the circuit where the ammeter can be inserted. Refer to Figure 13-6 for ammeter placement information. After the measurement is performed, make sure that every jumper is placed back in its place.

This concludes the procedure for this experiment.

Table 13-1

Predicted Current Measured Current V1=5.0V *1.2k Ω 560 Ω 470 Ω Total Current* ***V1=11.0V*** *1.2k Ω 560 Ω 470 Ω Total Current*

Questions

Perform the error calculation for all entries in Table 13-1. Comment on whether the errors are acceptable.

2. What general statement can be made about total current through a parallel circuit?

3. Explain what happens to total current as parallel branches are added and removed.

Bonus:

4. Using Ohm's law, demonstrate what a short across any parallel branch will do to the total circuit current. Make a general statement about a short circuit across any parallel branch.

14

Power in Parallel

Problem Statement

Electric heaters and light bulbs are rated in watts (W) and the motors are rated in horsepower (or watts): both are the units of power. The greater the power rating on the heater or the light bulb, the more heat or light is produced in a given time. Understanding power dissipation concepts helps us construct circuits that perform close to their rated capacity but not exceeding it. Consider the following challenge. You need to construct an electrical circuit that will provide power to electrical heaters. You need to be concerned with the maximum current carrying capacity of each circuit. Understanding how power dissipates in a parallel circuit will help you solve this challenge.

Concept

The characteristics of power in a parallel circuit are much the same as those in a series circuit. The power may be determined a number of ways:

$$P=IV, \qquad P=I^2 R, \qquad P=V^2 /R$$

The total power in a parallel circuit is equal to the sum of all individual power dissipations. Power is not measured; it is determined through calculation. Total power may also be determined using the various power equations. In order to calculate total power, total quantities must be used in the calculations, i.e., total voltage, total current, and total resistance. Whether the total power is determined using one of the equations for power or by adding the individual power dissipations, either method should yield a similar result.

Learning Objectives

Upon completion of this lab, the student will be able to:

1. Configure a circuit using eTCB schematic drawing and eTCB board.
2. Given any two electrical parameters voltage, current, or resistance, determine the power dissipation of an electrical load in parallel circuit.
3. Given any two electrical parameters voltage, current, or resistance, determine the total power dissipation in parallel circuit.

Pre-Lab Procedure

1. Use all three different power equations to predict the power dissipated in **each resistor** (Figure 14-1). Perform this calculation for supply voltage **V1=7.0V**. Record your results in Table 14-1.

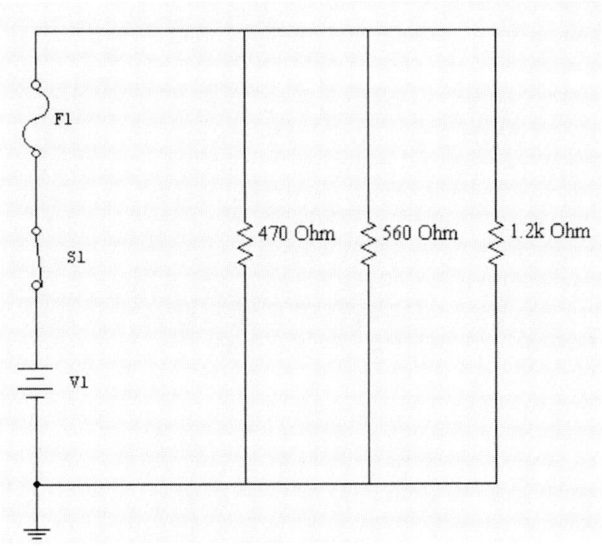

Figure 14-1

2. Use all three different power equations to predict the **total power** dissipated in the parallel circuit (Figure 14-1). Perform this calculation for supply voltage **V1=7.0V**. Record your results in Table 14-1.

3. Repeat Pre-Lab Procedure #1 and #2 for supply voltage **V1=11.0V**. Record your results in Table 14-2.

Lab Procedure

Before proceeding further, make sure that your eTCB is working correctly. Using eTCB Diagnostic VI, perform a self-diagnostic of eTCB circuitry.

Turn the switch S1 and S2 OFF in preparation for the lab.

Setup

Step #1 Using the eTCB Configuration VI, (DC module), first reset all relays and then configure the circuit in Figure 14-2.

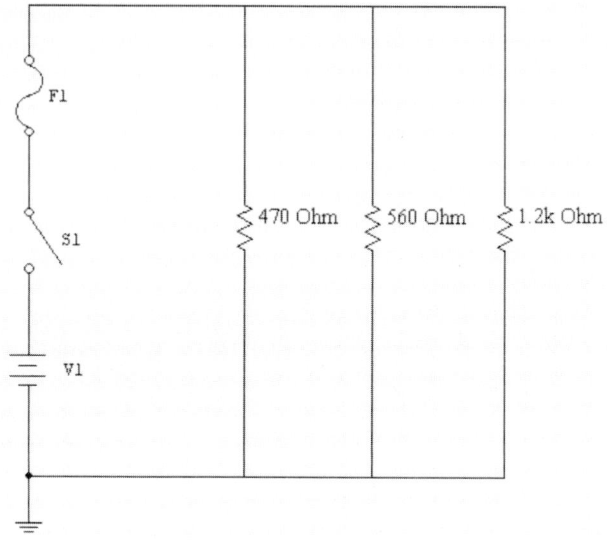

Figure 14-2

You will be asked to measure a DC current in this experiment, and as part of the DMM setup you need to "Null" the ammeter. This will minimize the error in DC current measurement. The "Nulling" procedure is very simple and needs to be performed only once per experiment. The procedure to "Null" the ammeter is as follows:

- Remove the test leads from the DMM terminals labeled "CURRENT".
- Open the Digital Multimeter VI and select DC Current measurement function.
- Click on the "Null" button.

Step #2 Using the Variable Power Supplies VI, set voltage source V1 to ***7.0V***. Keep in mind that this setting is just a good indication of what the voltage of V1 is set to. Variable power supplies often need to be adjusted (tuned) in order to set their output voltage precisely. Keep the variable power supplies VI ***open***; if you close it, all the variable voltage supplies will reset to 0V.

Step #3 Physically turn switch S1 on the eTCB board "ON". This will complete your parallel circuit. To verify that the switch is ON (closed) and that the fuse is not blown, use the Digital Multimeter VI to measure voltage at TP1 with respect to ground. Adjust the output of the variable power supplies, if necessary.

Discovery

Step #4 Using the Digital Multimeter VI, measure the current through the **1.2k Ω** resistor. Record this reading in Table 14-1 (column – Measured). To perform this measurement, you need to insert the ammeter in series with the **1.2K Ω** resistor. Removing the jumper **J4** creates a break in the circuit where the ammeter can be inserted. Refer to Figure 14-3 for ammeter placement information. After the measurement is performed, make sure that every jumper is placed back in its place.

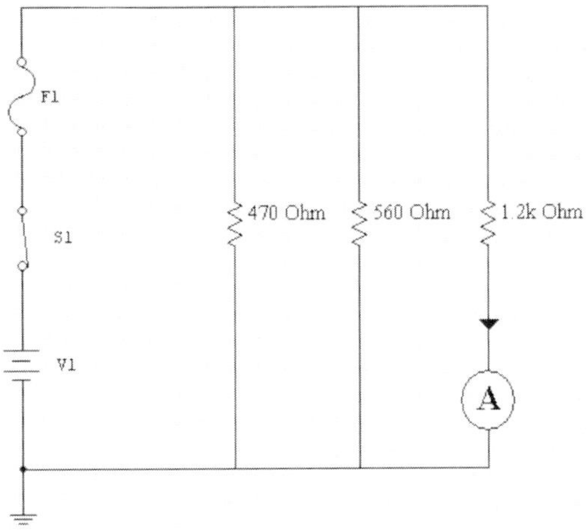

Figure 14-3

Step #5 Using the Digital Multimeter VI, measure the current through the **560 Ω** resistor. Record this reading in Table 14-1 (column – Measured). To perform this measuremen, you need to insert the ammeter in series with the **560 Ω** resistor. Removing the jumper **J3** creates a break in the circuit where the ammeter can be inserted. Refer to Figure 14-4 for ammeter placement information. After the measurement is performed, make sure that every jumper is placed back in its place.

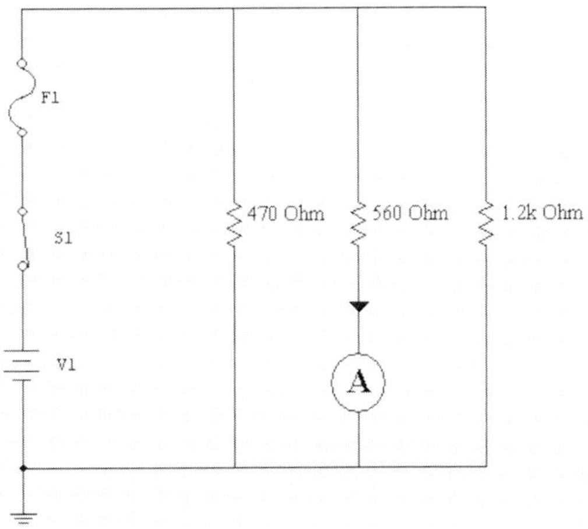

Figure 14-4

Step #6 Using the Digital Multimeter VI, measure the current through the **470 Ω** resistor. Record this reading in Table 14-1 (column – Measured). To perform this measurement, you need to insert the ammeter in series with the **470 Ω** resistor. Removing the jumper **J2** creates a break in the circuit where the ammeter can be inserted. Refer to Figure 14-5 for ammeter placement information. After the measurement is performed, make sure that every jumper is placed back in its place.

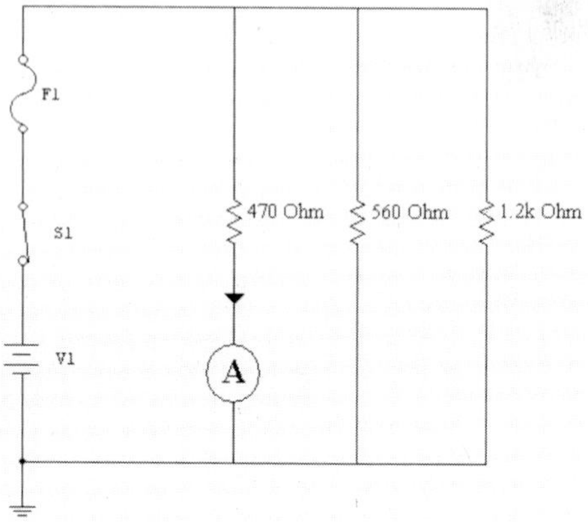

Figure 14-5

Step #7 Using the Digital Multimeter VI, measure the total current into the parallel circuit. Record this reading in Table 14-1 (column – Measured). Removing the jumper **I1** creates a break in the circuit where the ammeter can be inserted. Refer to Figure 14-6 for ammeter placement information. After the measurement is performed, make sure that every jumper is placed back in its place.

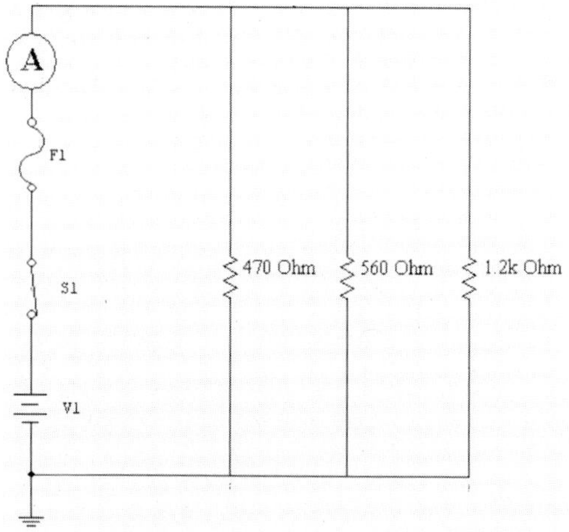

Figure 14-6

Note: We are now going to examine the effects of different supply voltage on the currents in the parallel circuit. We are going to use the same circuit diagram, but rather than setting V1=5.0V, we are going to set it at ***V1=11.0V.***

Step #8 Using the Variable Power Supplies VI, set voltage source V1 to ***11.0V.***
Using the Digital Multimeter VI, measure voltage at TP1 with respect to ground. Adjust the output of the variable power supplies, if necessary.

Step #9 Using the Digital Multimeter VI, measure the current through the **1.2k Ω** resistor. Record this reading in Table 14-2 (column – Measured). To perform this measurement, you need to insert the ammeter in series with the **1.2K Ω** resistor. Removing the jumper **J4** creates a break in the circuit where the ammeter can be inserted. Refer to Figure 14-3 for ammeter placement information. After the measurement is performed, make sure that every jumper is placed back in its place.

Step #10 Using the Digital Multimeter VI, measure the current through the **560 Ω** resistor. Record this reading in Table 14-2 (column – Measured). To perform this measurement, you need to insert the ammeter in series with the **560 Ω** resistor. Removing the jumper **J3** creates a break in the circuit where the ammeter can be inserted. Refer to Figure 14-4 for ammeter placement information. After the measurement is performed, make sure that every jumper is placed back in its place.

Step #11 Using the Digital Multimeter VI, measure the current through the **470 Ω** resistor. Record this reading in Table 14-2 (column – Measured). To perform this measurement you need to insert the ammeter in series with the **470 Ω** resistor. Removing the jumper **J2** creates a break in the circuit where the ammeter can be inserted. Refer to Figure 14-5 for ammeter placement information. After the measurement is performed, make sure that every jumper is placed back in its place.

Step #12 Using the Digital Multimeter VI, measure the total current into the parallel circuit. Record this reading in Table 14-2 (column – Measured). Removing the jumper **I1** creates a break in the circuit where the ammeter can be inserted. Refer to Figure 14-6 for ammeter placement information. After the measurement is performed, make sure that every jumper is placed back in its place.

This concludes the procedure for this experiment.

Table 14-1

	Predicted	***Measured***
Supply Voltage (V1)	**7.0V**	
Current		
Total Power (VI)		
Total Power (V^2/R)		
Total Power (I^2R)		
1.2k Ω Resistor		
Resistor Current		
Resistor Power (VI)		
Resistor Power (V^2/R)		
Resistor Power (I^2R)		
560 Ω Resistor		
Resistor Current		
Resistor Power (VI)		
Resistor Power (V^2/R)		
Resistor Power (I^2R)		
470 Ω Resistor		
Resistor Current		
Resistor Power (VI)		
Resistor Power (V^2/R)		
Resistor Power (I^2R)		

Table 14-2

	Predicted	*Measured*
Supply Voltage (VI)	**11.0V**	
Current		
Total Power (VI)		
Total Power (V^2/R)		
Total Power (I^2R)		
	1.2k Ω Resistor	
Resistor Current		
Resistor Power (VI)		
Resistor Power (V^2/R)		
Resistor Power (I^2R)		
	560 Ω Resistor	
Resistor Current		
Resistor Power (VI)		
Resistor Power (V^2/R)		
Resistor Power (I^2R)		
	470 Ω Resistor	
Resistor Current		
Resistor Power (VI)		
Resistor Power (V^2/R)		
Resistor Power (I^2R)		

Questions

1. Perform the error calculation for all entries in Table 14-1 and Table 14-2. Comment on whether the errors are acceptable.

2. Since voltage is the same in a parallel circuit, what determines the power dissipation in each of the resistive loads?

3. How does the total power in a parallel circuit relate the individual powers dissipated in the resistive loads?

Bonus:

4. You have a number of 500W electric heaters available to provide heat for a large room in a building. This room requires at least 8000W of power to be adequately heated. Each circuit supplying power to the heaters has a current caring capacity of 20A, Figure 14-7. What is the minimum number of circuits that must be installed in this room?

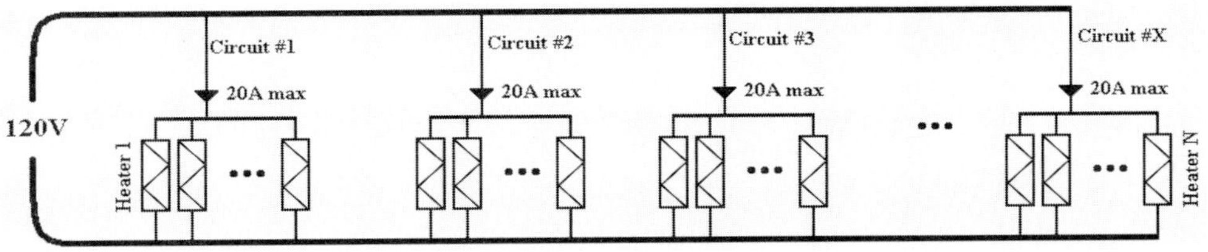

Figure 14-7

15

Resistance in Series-Parallel

Problem Statement

At times a resistance sensitive circuit may call for a resistance value that is not a standard value. Understanding how equivalent resistance is determined in series-parallel circuits will help in building a circuit with multiple standard value resistors, which together will offer the desired equivalent resistance.

Concept

Most complex circuits can be reduced to a simple series circuit having a single load and a single voltage source. Total resistance in a series-parallel circuit may be reduced to a single series resistance. Once the direction of current flow is determined, begin the analysis at the last current division. The current flow theory used is irrelevant.

At the last current division, reduce the combination of resistors to a single resistance. Once this is determined, redraw the circuit with the new equivalent resistance. Continue this process until all of the resistances have been combined into a single load resistance. This process combines the concepts of series resistances and parallel resistances. This is demonstrated in Figure 15-1.

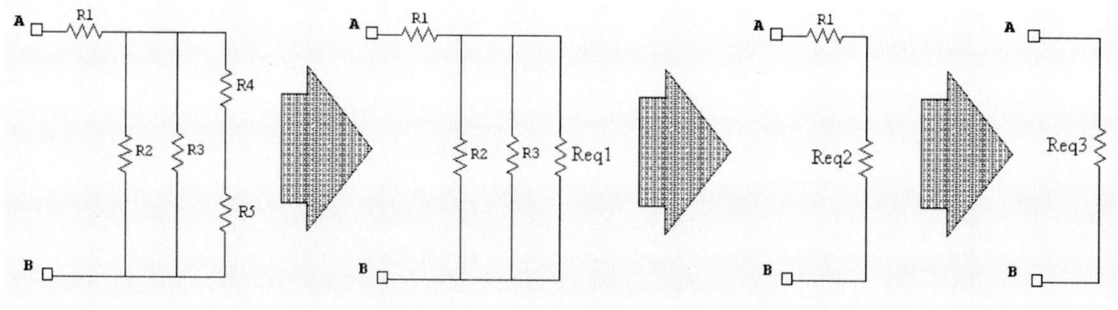

Figure 15-1

Req1 = R4 + R5
Req2 = (R4 + R5) // R2 // R3
Req3 = R1 + (R4 + R5) // R2 // R3

Learning Objectives

Upon completion of this lab, the student will be able to:

1. Configure a circuit using eTCB schematic drawing and eTCB board.
2. Predict and measure the resistance in a series-parallel circuit.

Pre-Lab Procedure

1. Given the circuit in Figure 15-2A and resistors color code in Table 15-1, determine the resistance of each of the resistors as well as the equivalent resistance seen between points A and B. Record the equivalent resistance result in Table 15-2.

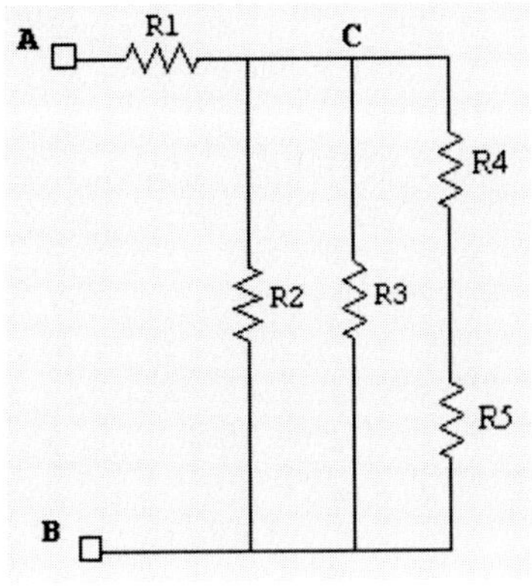

Figure 15-2A

2. Repeat Pre-Lab Procedure #1 for the series-parallel circuit with series resistor branch removed, as shown in Figure 15-2B.

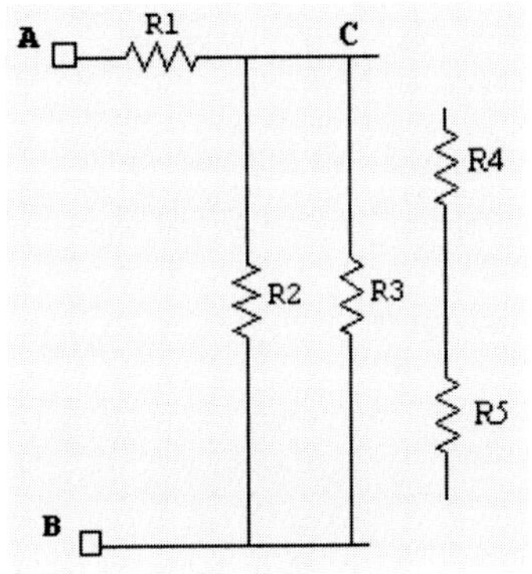

Figure 15-2B

3. Repeat Pre-Lab Procedure #1 for the series-parallel circuit with R1 resistor shorted, as shown in Figure 15-2C.

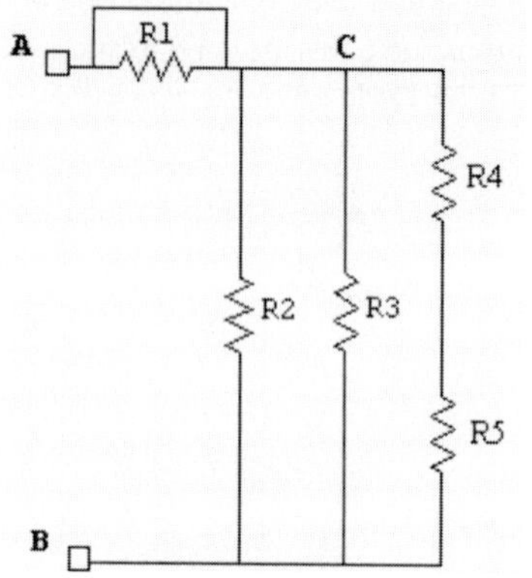

Figure 15-2C

Table 15-1

	First Band	***Second Band***	***Third Band***	***Forth Band***	***Rated Resistance***	***Tolerance***
R1	*Red*	*Red*	*Brown*	*Gold*		
R2	*Green*	*Blue*	*Brown*	*Gold*		
R3	*Brown*	*Red*	*Red*	*Gold*		
R4	*Brown*	*Black*	*Red*	*Gold*		
R5	*Green*	*Blue*	*Brown*	*Gold*		

Lab Procedure

Before proceeding further. make sure that your eTCB is working correctly. Using eTCB Diagnostic VI, perform a self-diagnostic of eTCB circuitry.

Turn the switch S1 and S2 OFF in preparation for the lab.

Setup

Step #1 Using the eTCB Configuration VI, (DC module), reset all relays and then configure the circuit in Figure 15-3.

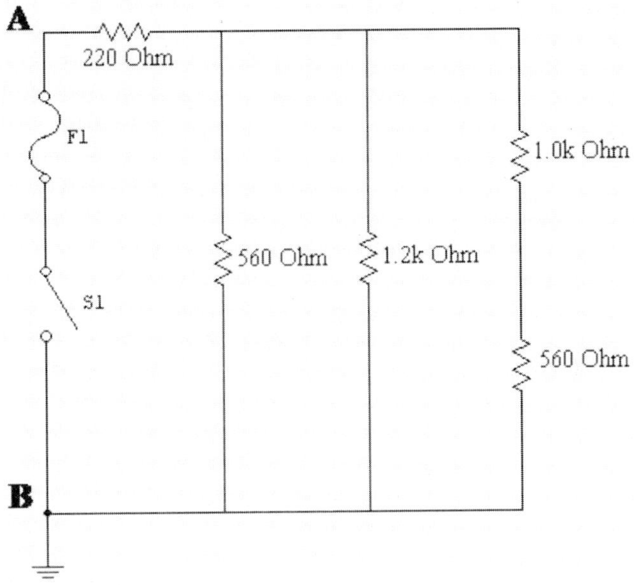

Figure 15-3

Note: Make sure that switch S1 remains "open" "OFF" and all the power supplies are completely isolated from any components on the eTCB board.

Discovery

Step #2 Using the Digital Multimeter VI, measure the equivalent resistance of the series-parallel circuit seen from points **A and B.** Record this reading in Table 15-2 (column – Measured).

Step #3 Using the Digital Multimeter VI, measure the equivalent resistance between points **C and B.** Refer to Figure 15-4 for the DMM placement information. Record the DMM reading in Table 15-2 (column – Measured).

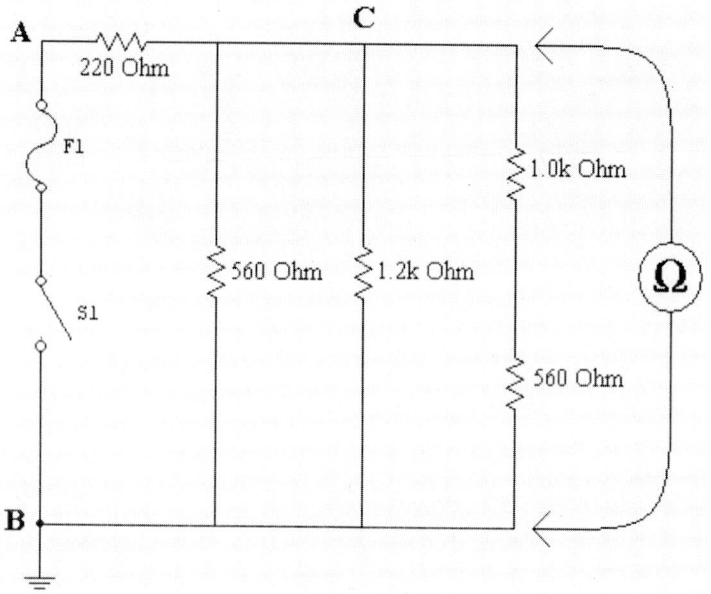

Figure 15-4

Step #4

In this step, we are going to measure the equivalent resistance between points **D and B.** To isolate this branch from the rest of the circuit, use the eTCB Configuration VI. Refer to Figure 15-5 for the DMM placement information. Using the Digital Multimeter VI, measure the equivalent resistance. Record the DMM reading in Table 15-2 (column – Measured).

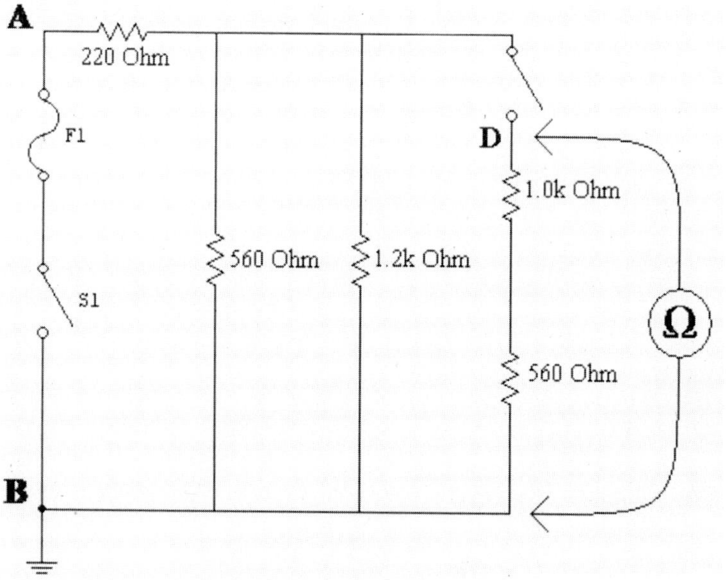

Figure 15-5

Step #5 Keep the same configuration as in Figure 15-5, effectively removing series resistors branch from the original circuit. Using the Digital Multimeter VI, measure the equivalent resistance of the series-parallel circuit seen from points **A and B.** Record this reading in Table 15-2 (column – Measured).

Step #6 Using the eTCB Configuration VI, (DC Module), configure the circuit in Figure 15-6. This circuit has 220 Ω resistor which is shorted.

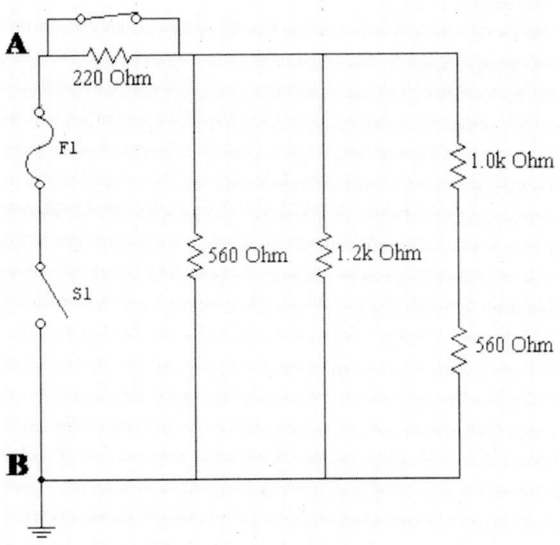

Figure 15-6

Note: Make sure that switch S1 remains "open" "OFF" and all the power supplies are completely isolated from any components on the eTCB board.

Step #7 Using the Digital Multimeter VI, measure the equivalent resistance of the series-parallel circuit seen from points **A and B.** Record this reading in Table 15-2 (column – Measured).

This concludes the procedure for this experiment.

Table 15-2

	Predicted	***Measured***
Resistance between **C and B**		
Resistance between **D and B**		
Equivalent Resistance		
Original circuit with series resistor branch removed		
Equivalent Resistance		
Original circuit with 220 Ω resistor shorted		
Equivalent Resistance		

Questions

1. Perform the error calculation for all entries in Table 15-2. Comment on whether the errors are acceptable.

2. How is total (equivalent) resistance in series-parallel circuits determined?

3. What effect does adding additional loads (resistors) in parallel have on the total resistance of the series-parallel circuits?

Bonus:

4. Given many of 2.2kΩ resistors, and 3.3kΩ resistors, construct a circuit that will have an equivalent resistance of 4.4kΩ.

16

Current in Series-Parallel

Concept

Series-parallel circuits possess the characteristics of both series and parallel circuits. The total current leaves the source and must return to the source. Any load in series with the source will have the total circuit current flowing through it. As parallel combinations of resistors are encountered, the current will divide and can be determined by current divider equations. The total current in the series-parallel circuit is determined using Ohm's law. All the rules of current flow learned in series and parallel circuits individually apply to the following series-parallel circuit.

Learning Objectives

Upon completion of this lab, the student will be able to:

1. Configure a circuit using eTCB schematic drawing and eTCB board.
2. Given voltage and resistance, predict the current in a series-parallel circuit.
3. Measure the current in multiple places in a series-parallel circuit.

Pre-Lab Procedure

1. Given the voltage V1=5.0V, predict current ***I*** in the circuit shown in Figure 16-1. Record your result in Table 16-1.

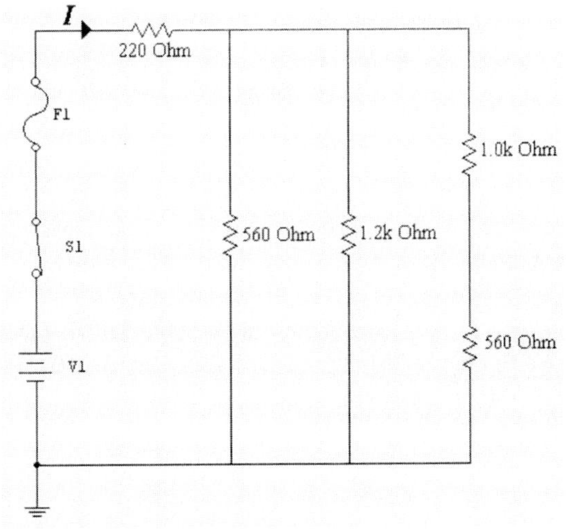

Figure 16-1

2. Repeat Pre-Lab Procedure 1 with the voltage V1=11.0V. Record your result in Table 16-1.

Lab Procedure

Before proceeding further, make sure that your eTCB is working correctly. Using eTCB Diagnostic VI, perform a self-diagnostic of eTCB circuitry.

Turn the switch S1 and S2 OFF in preparation for the lab.

Setup

Step #1 Using the eTCB Configuration VI, (DC Module), reset all relays and then configure the circuit in Figure 16-2.

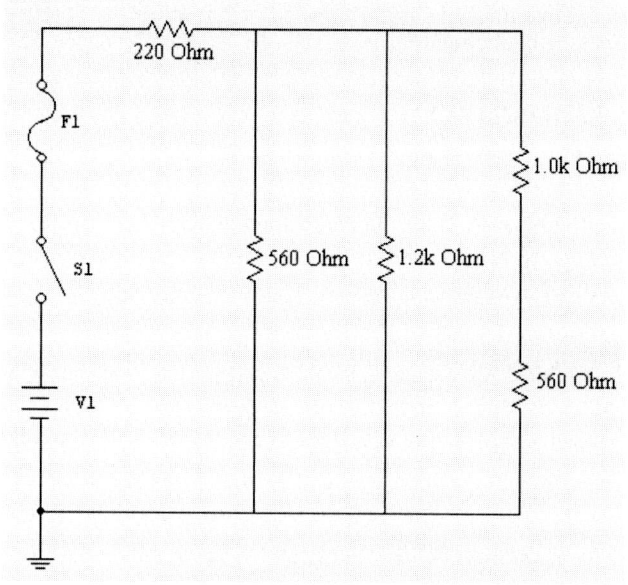

Figure 16-2

You will be asked to measure a DC current in this experiment, and as part of the DMM setup you need to "Null" the ammeter. This will minimize the error in DC current measurement. The "Nulling" procedure is very simple and needs to be performed only once per experiment. The procedure to "Null" the ammeter is as follows:

- Remove the test leads from the DMM terminals labeled "CURRENT".
- Open the Digital Multimeter VI and select DC Current measurement function.
- Click on the "Null" button.

Step #2 Using the Variable Power Supplies VI, set voltage source V1 to ***5.0V.*** Keep in mind that this setting is just a good indication of what the voltage of V1 is set to. variable power Supplies often need to be adjusted (tuned) in order to set their output voltage precisely. Keep the variable power supplies VI ***open***; if you close it, all the variable voltage supplies will reset to 0V.

Step #3 Physically turn switch S1 on the eTCB board "ON". This will complete your series-parallel circuit. To verify that the switch is ON (closed) and that the fuse is not blown, use the Digital Multimeter VI to measure voltage at TP1 with respect to ground. Adjust the output of the variable power supplies, if necessary.

Discovery

Step #4 Using the Digital Multimeter VI, measure the current through the series branch of **1.0k Ω** and **560 Ω** resistor. Record this reading in Table 16-1 (column – Measured). To perform this measurement, you need to insert the ammeter in series with the **560 Ω** resistor. Removing the jumper **J7** creates a break in the circuit where the ammeter can be inserted. Refer to Figure 16-3 for ammeter placement information. After the measurement is performed, make sure that every jumper is placed back in its place.

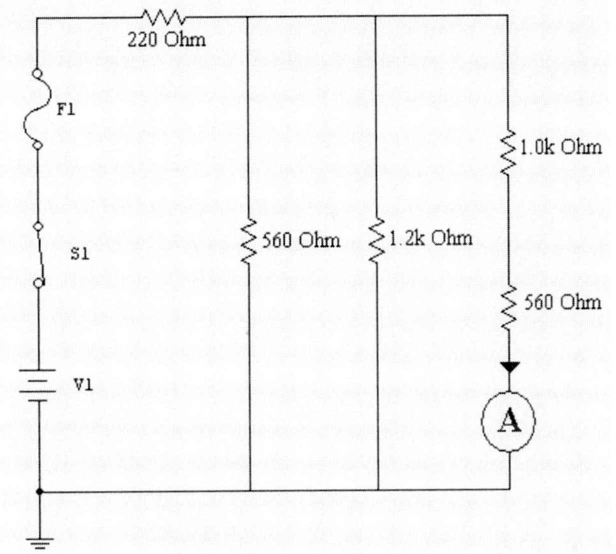

Figure 16-3

Step #5 Using the Digital Multimeter VI, measure the current through the **1.2k Ω** resistor. Record this reading in Table 16-1 (column – Measured). To perform this measurement, you need to insert the ammeter in series with the **1.2k Ω** resistor. Removing the jumper **J4** creates a break in the circuit where the ammeter can be inserted. Refer to Figure 16-4 for ammeter placement information. After the measurement is performed, make sure that every jumper is placed back in its place.

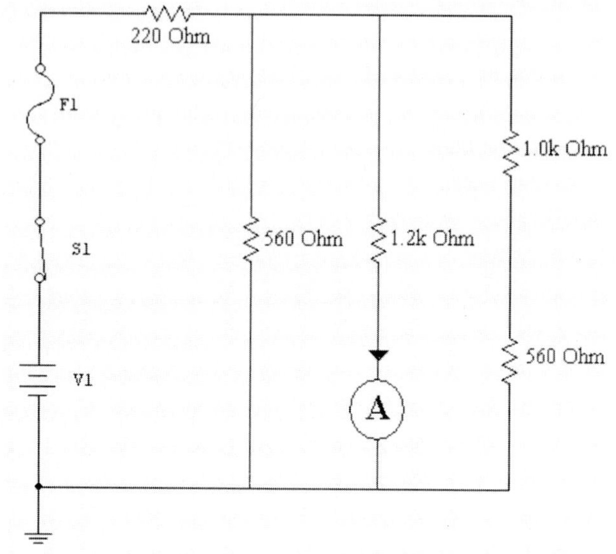

Figure 16-4

Step #6 Using the Digital Multimeter VI, measure the current through the **560 Ω** resistor. Record this reading in Table 16-1 (column – Measured). To perform this measurement, you need to insert the ammeter in series with the **560 Ω** resistor. Removing the jumper **J3** creates a break in the circuit where the ammeter can be inserted. Refer to Figure 16-5 for ammeter placement information. After the measurement is performed, make sure that every jumper is placed back in its place.

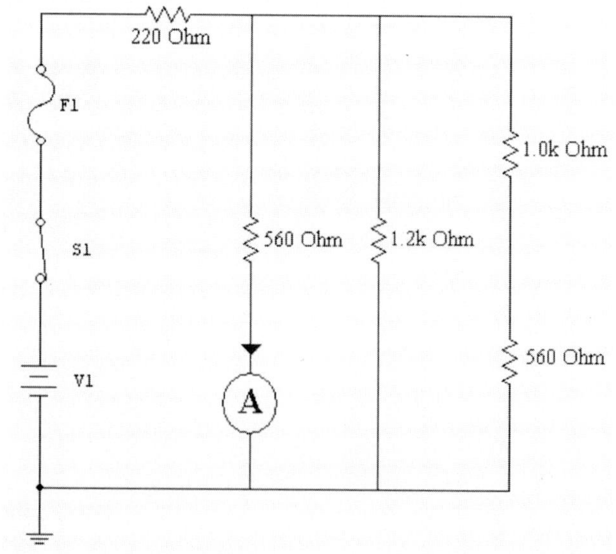

Figure 16-5

Step #7 Using the Digital Multimeter VI, measure the total current into the series-parallel circuit. Record this reading in Table 16-1 (column – Measured). Removing the jumper **I1** creates a break in the circuit where the ammeter can be inserted. Refer to Figure 16-6 for ammeter placement information. After the measurement is performed, make sure that every jumper is placed back in its place.

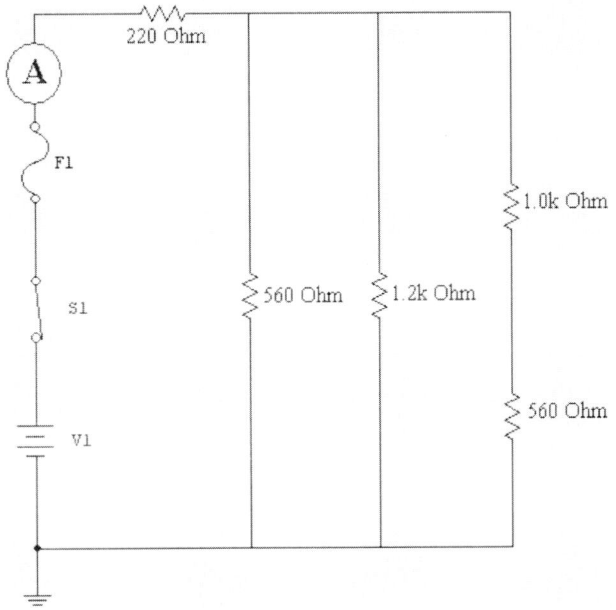

Figure 16-6

*Note: We are now going to examine the effects of different supply voltage on the currents in the series-parallel circuit. We are going to use the same circuit diagram, but rather than setting V1=5.0V, we are going to set it at **V1=11.0V.***

Step #8 Using the Variable Power Supplies VI, set voltage source V1 to ***11.0V.***
Using the Digital Multimeter VI, measure voltage at TP1 with respect to ground. Adjust the output of the variable power supplies, if necessary.

Step #9 Using the Digital Multimeter VI, measure the current through the series branch of **1.0k Ω** and **560 Ω** resistor. Record this reading in Table 16-1 (column – Measured). To perform this measurement, you need to insert the ammeter in series with the **560 Ω** resistor. Removing the jumper **J7** creates a break in the circuit where the ammeter can be inserted. Refer to Figure 16-3 for ammeter placement information. After the measurement is performed, make sure that every jumper is placed back in its place.

Step #10 Using the Digital Multimeter VI, measure the current through the **1.2k Ω** resistor. Record this reading in Table 16-1 (column – Measured). To perform this measurement, you need to insert the ammeter in series with the **1.2k Ω** resistor. Removing the jumper **J4** creates a break in the circuit where the ammeter can be inserted. Refer to Figure 16-4 for ammeter placement information. After the measurement is performed, make sure that every jumper is placed back in its place.

Step #11 Using the Digital Multimeter VI, measure the current through the **560 Ω** resistor. Record this reading in Table 16-1 (column – Measured). To perform this measurement, you need to insert the ammeter in series with the **560 Ω** resistor. Removing the jumper **J3** creates a break in the circuit where the ammeter can be inserted. Refer to Figure 16-5 for ammeter placement information. After the measurement is performed, make sure that every jumper is placed back in its place.

Step #12 Using the Digital Multimeter VI, measure the total current into the series-parallel circuit. Record this reading in Table 16-1 (column – Measured). Removing the jumper **I1** creates a break in the circuit where the ammeter can be inserted. Refer to Figure 16-6 for ammeter placement information. After the measurement is performed, make sure that every jumper is placed back in its place.

This concludes the procedure for this experiment.

Table 16-1

	Predicted Current	Measured Current
V1=5.0V		
1.0k Ω in series with 560 Ω		
1.2k Ω		
560 Ω		
Total Current		
V1=11.0V		
1.0k Ω in series with 560 Ω		
1.2k Ω		
560 Ω		
Total Current		

Questions

1. Perform the error calculation for all entries in Table 16-1. Comment on whether the errors are acceptable.

2. How are the currents through each resistive load in a series-parallel circuit related to the resistance of each load?

3. The current leaving a parallel branch in a series-parallel circuit must equal what value?

17

Voltage in Series-Parallel

Concept

Series-parallel circuits possess characteristics of both series and parallel circuits; voltage in these types of circuits will be determined accordingly. Voltage across parallel branches will be the same across all branches. Voltage developed across series loads is directly proportional to the resistance of the load and the current through the load.

The voltage across any load may be determined using Ohm's law. Once the current and resistance of the load are determined, they are multiplied together to determine the voltage across the load. All series voltages are additive, all voltages in parallel are the same value.

Learning Objectives

Upon completion of this lab, the student will be able to:

1. Configure a circuit using eTCB schematic drawing and eTCB board.
2. Given supply voltage value, predict voltage drops on each resistor in a series-parallel circuit.
3. Measure the voltage in a series-parallel circuit.

Pre-Lab Procedure

1. Given the voltage V1=7.0V, predict the voltage drops $\mathbf{V_{DB}}$, $\mathbf{V_{CB}}$, and $\mathbf{V_{AC}}$ in the series-parallel circuit shown in Figure 17-1. Record your result in Table 17-1.

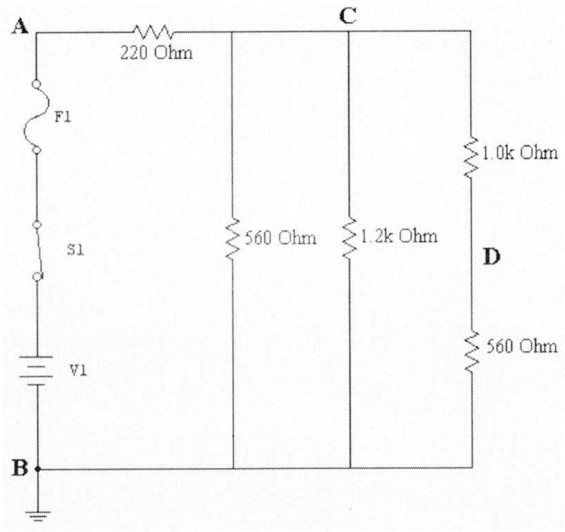

Figure 17-1

2. Repeat Pre-Lab Procedure 1 with the voltage V1=11.0V. Record your result in Table 17-1.

Lab Procedure

Before proceeding further, make sure that your eTCB is working correctly. Using eTCB Diagnostic VI, perform a self-diagnostic of eTCB circuitry.

Turn the switch S1 and S2 OFF in preparation for the lab.

Setup

Step #1 Using the eTCB Configuration VI, (DC module), reset all relays and then configure the circuit in Figure 17-2.

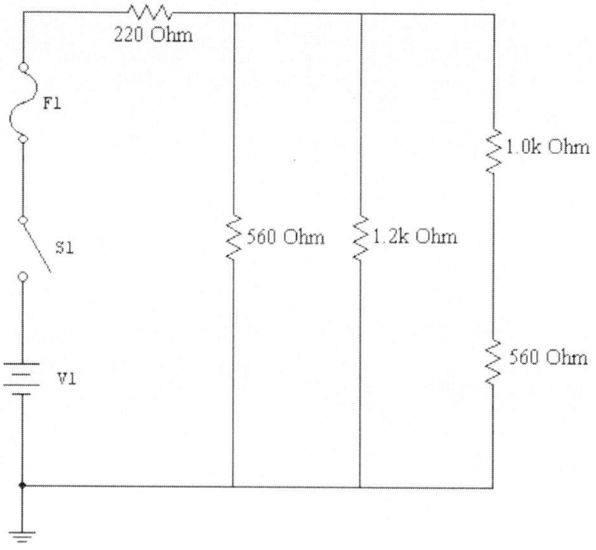

Figure 17-2

Step #2 Using the Variable Power Supplies VI, set voltage source V1 to ***7.0V.*** Keep in mind that this setting is just a good indication of what the voltage of V1 is set to. Variable power supplies often need to be adjusted (tuned) in order to set their output voltage precisely. Keep the Variable Power Supplies VI ***open***; if you close it, all the variable voltage supplies will reset to 0V.

Step #3 Physically turn switch S1 on the eTCB board "ON". This will complete your parallel circuit. To verify that the switch is ON (closed) and that the fuse is not blown, use the Digital Multimeter VI to measure voltage at TP1 with respect to ground. Adjust the output of the variable power wupplies, if necessary.

Discovery

Step #4 Using the Digital Multimeter VI, measure the voltage drop V_{DB}. Record this reading in Table 17-1 (column – Measured). Refer to Figure 17-3 for the DMM placement information.

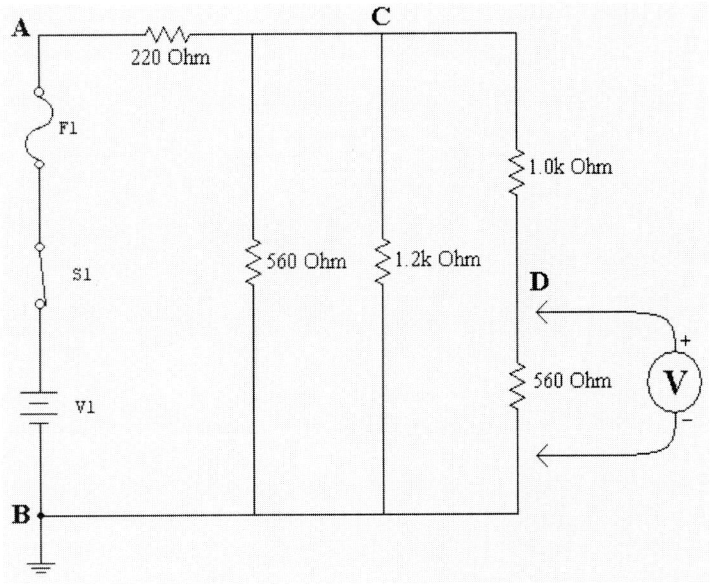

Figure 17-3

Step #5 Using the Digital Multimeter VI, measure the voltage drop V_{CB}. Record this reading in Table 17-1 (column – Measured).

Step #6 Using the Digital Multimeter VI, measure the voltage drop V_{AC}. Record this reading in Table 17-1 (column – Measured).

*Note: We are now going to examine the effects of different supply voltage on the voltage drops across the individual resistors in the series-parallel circuit. We are going to use the same circuit diagram, but rather than setting V1=7.0V, we are going to set it at **V1=11.0V.***

Step #7 Using the Variable Power Supplies VI, set voltage source V1 to ***11.0.V.***
Using the Digital Multimeter VI, measure voltage at **TP1** with respect to ground. Adjust the output of the variable power supplies, if necessary.

Step #8 Using the Digital Multimeter VI, measure the voltage drop V_{DB}. Record this reading in Table 17-1 (column – Measured).

Step #9 Using the Digital Multimeter VI, measure the voltage drop V_{CB}. Record this reading in Table 17-1 (column – Measured).

Step #10 Using the Digital Multimeter VI, measure the voltage drop V_{AC}. Record this reading in Table 17-1 (column – Measured).

This concludes the procedure for this experiment.

Table 17-1

	Predicted Voltage	*Measured Voltage*
V1=7.0V		
V_{DB}		
V_{CB}		
V_{AC}		
V1=11.0V		
V_{DB}		
V_{CB}		
V_{AC}		

Questions

1. Perform the error calculation for all entries in Table 17-1. Comment on whether the errors are acceptable.

2. When are voltages equal in a series-parallel circuit?

3. When can the voltage divider equation be used to determine voltage drop across a resistor in a series-parallel circuit?

4. When can the current divider equation be used to determine current through a resistor in a series-parallel circuit?

5. When the circuit is reduced to an equivalent series circuit, what must the algebraic sums of the voltages equal?

18

Wheatstone Bridge

Problem Statement

In this experiment, we will examine the behavior of a bridge circuit. Consider a scenario where you need to measure a resistance of a pressure sensor. This particular sensor reflects a change in pressure as change in resistance. Our challenge is to measure resistance of the sensor (without the use of an ohmmeter). Furthermore, by measuring voltage, we want to be able to determine if the resistance of a sensor is greater or less than the one at a specific set point. In this experiment, you will learn how the Wheatstone bridge circuit can be used to address the above challenge.

Concept

The Wheatstone bridge is an application of a series-parallel circuit. It may be the most common circuit application in process control today. The bridge can compare a measurable quantity to some predetermined reference point. From a difference in voltage it may be determined if the measured quantity is greater than or less than the set-point value represented by the reference point.

Most of the time, what is being measured is a non-electrical quantity. The bridge, with the application of a transducer, describes changes in a non-electrical phenomenon as changes in voltage. A transducer is a device that converts a non-electrical quantity into a related electrical quantity. Examples of this are pressure-sensitive resistors, light-sensitive resistors, and temperature-sensitive resistors.

Bridges are balanced when there is no difference of potential between the center reference points, as shown in Figure 18-1.

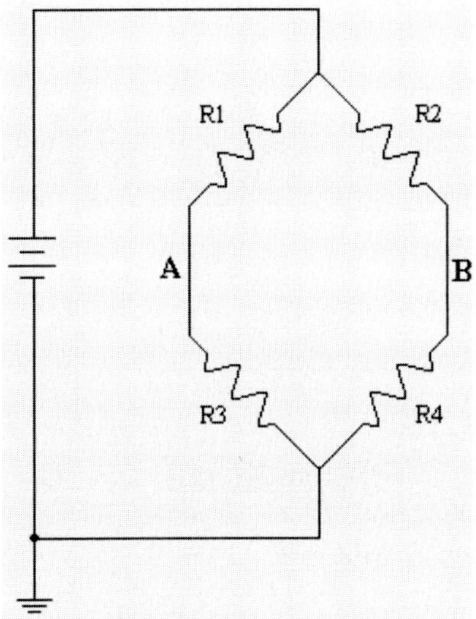

Figure 18-1

The bridge is balanced when the following ratios of resistors exist:

R1/R2=R3/R4 or R1/R3=R2/R4

These ratios are subject to algebraic manipulation. Using these ratios the value of an unknown resistor may be determined, making an ohmmeter from a bridge. Assuming R_2 to be R unknown:

R1R4/R3=Runknown

Learning Objectives

Upon completion of this lab, the student will be able to:

1. Configure a circuit using eTCB schematic drawing and eTCB board.
2. Given supply voltage value, predict voltage drops on each resistor in a Wheatstone Bridge circuit.
3. Measure voltage in a Wheatstone Bridge circuit.
4. Determine if a Wheatstone Bridge is balanced.

Pre-Lab Procedure

1. Given the voltage V1=10.0V and resistance Rp = 1.9k _, predict the voltage $\mathbf{V_A}$, $\mathbf{V_B}$, and $\mathbf{V_{AB}}$ in the Wheatstone Bridge circuit shown in Figure 18-2. Record your result in Table 18-1.

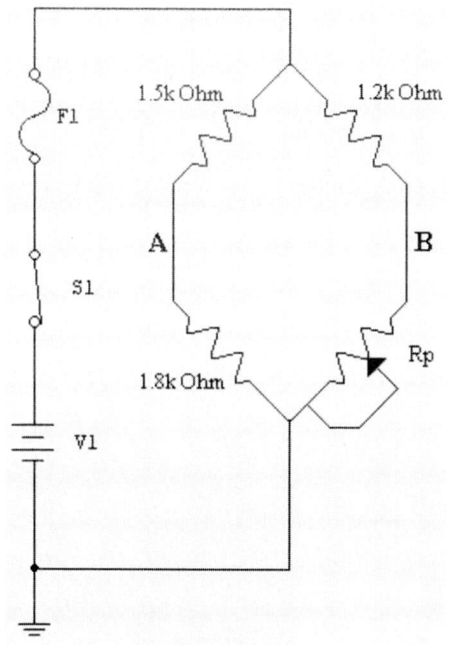

Figure 18-2

2. Repeat Pre-Lab procedure #1 with resistance Rp = 800 _.

3. What is the resistance value of a potentiometer Rp, in order to balance the Wheatstone Bridge circuit in Figure 18-2. Record your result in Table 18-1.

Lab Procedure

Before proceeding further, make sure that your eTCB is working correctly. Using eTCB Diagnostic VI, perform a self-diagnostic of eTCB circuitry.

Turn the switch S1 and S2 OFF in preparation for the lab.

.

Setup

Step #1 Using the eTCB Configuration VI, (DC module), reset all relays and then configure the circuit in Figure 18-3.
Using the Digital Multimeter VI, measure resistance of Rp while adjusting it to **1.9k _.**

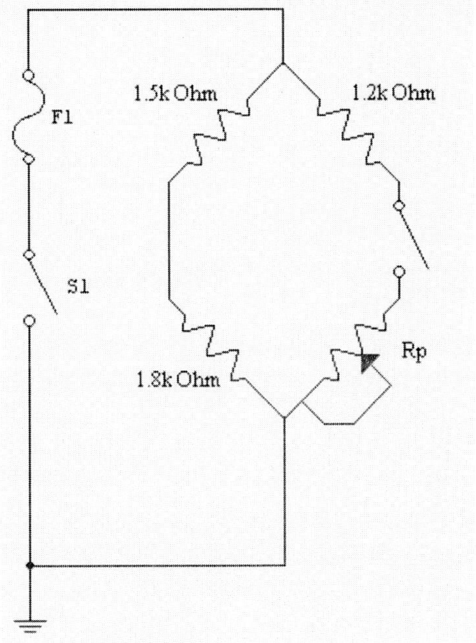

Figure 18-3

Step #2 Using the eTCB Configuration VI, (DC module), configure the circuit in Figure 18-4

Using the Variable Power Supplies VI, set voltage source V1 to ***10.0V.*** Keep in mind that this setting is just a good indication of what the voltage of V1 is set to. Variable power supplies often need to be adjusted (tuned) in order to set their output voltage precisely. Keep the Variable Power Supplies VI ***open***; if you close it, all the variable voltage supplies will reset to 0V.

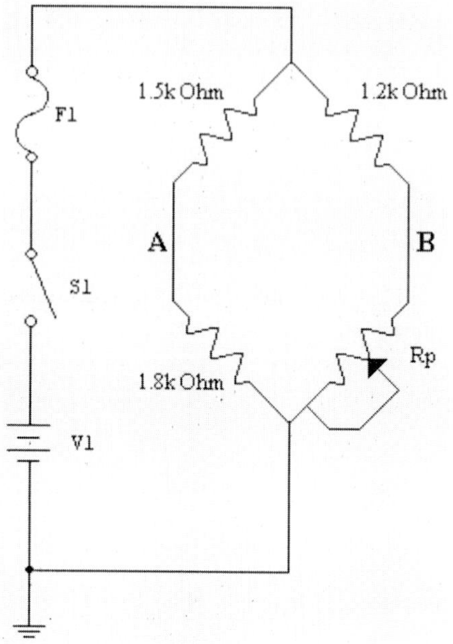

Figure 18-4

Step #3 Using the Digital Multimeter VI, measure V1 (V_{Source}) voltage with respect to ground. Adjust the output of the variable power supplies, if necessary.

Step #4 Physically turn switch S1 on the eTCB board "ON". This will complete your circuit. To verify that the switch is "ON" (closed) and that the fuse is not blown, use the Digital Multimeter VI to measure voltage at TP1 with respect to ground.

Discovery

Step #5 Using the Digital Multimeter VI, measure voltage $\mathbf{V_A}$. Record this reading in Table 18-1 (column – Measured).

Note: this voltage measurement is with respect to ground.

Step #6 Using the Digital Multimeter VI, measure voltage $\mathbf{V_B}$. Record this reading in Table 18-1 (column – Measured).

Note: this voltage measurement is with respect to ground.

Step #7 Using the Digital Multimeter VI, measure voltage $\mathbf{V_{AB}}$. Record this reading in Table 18-1 (column – Measured).

*Note: We are now going to examine the effects of different resistance or Rp on the voltages at point A and B of the Wheatstone Bridge circuit. We are going to use the same circuit diagram, but rather than setting Rp=1.9k _, we are going to set it at **Rp = 800 _**.*

Step #8 Using the eTCB Configuration VI, (DC module), configure the circuit in Figure 18-3. Using the Digital Multimeter VI, measure resistance of Rp while adjusting it to **800 _.**

Step #9 Using the eTCB Configuration VI, (DC module), configure the circuit in Figure 18-4.

Step #10 Using the Digital Multimeter VI, measure voltage $\mathbf{V_A}$. Record this reading in Table 18-1 (column – Measured).

Note: this voltage measurement is with respect to ground.

Step #11 Using the Digital Multimeter VI, measure voltage $\mathbf{V_B}$. Record this reading in Table 18-1 (column – Measured).

Note: this voltage measurement is with respect to ground.

Step #12 Using the Digital Multimeter VI, measure voltage $\mathbf{V_{AB}}$. Record this reading in Table 18-1 (column – Measured).

Note: We are now going to balance the bridge.

Step #13 Continue using Digital Multimeter VI to measure voltage $\mathbf{V_{AB}}$ while adjusting resistance Rp till voltage $\mathbf{V_{AB}}$ is within 20mV of **0V.**

Step #14 Using the eTCB Configuration VI, (DC module), configure the circuit in Figure 18-3. Using the Digital Multimeter VI, measure resistance of Rp. Record this reading in Table 18-1 (column – Measured).

This concludes the procedure for this experiment.

Table 18-1

	Predicted	***Measured***
Rp=1.9k _		
V_A		
V_B		
V_{AB}		
Rp=800 _		
V_A		
V_B		
V_{AB}		
Resistance $\mathbf{R_P}$ necessary to balance the Bridge		
R_P		

Questions

1. What is the voltage between A and B when a bridge is balanced?

2. Would placing a ground reference point in the bridge circuit change the direction of current flow between A and B or change the polarity of A or B?

3. Describe the ratio of resistors that make up a bridge circuit when the bridge is balanced.

4. What is the primary use for a bridge circuit?

Bonus:

5. Given the voltage VAB=0V and R1=1.0k _, R3=1.2k _, and R4=1.8k _, determine the resistance of resistor R2.

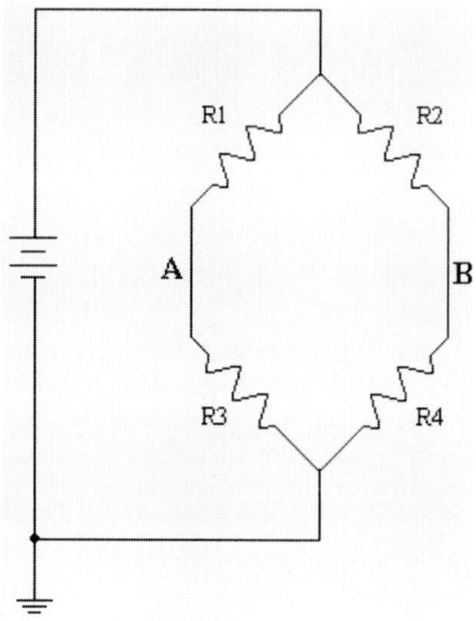

Figure 18-5

19

Thevenin's Equivalent Circuit

Problem Statement

It is not uncommon for series-parallel circuits to have variable loads. Each time the load changes, the entire circuit would require complete analysis in order to determine the operating characteristics of the load. Consider a scenario where you do not know how the circuit is built: you only have access to its power source and the terminals where a load is connected. Your task is to determine the value of the load resistor and its dissipated power, at the condition of maximum power transferred.

Thevenin's theorem analyzes the circuits from the point of view of the loads. The circuits are reduced to single voltage source with internal resistances. This approach allows an application of simplified circuit analysis to determine operating characteristics of the load.

Concept

In the case of a Thevenin's equivalent circuit, an equivalent resistance is placed in series with an equivalent voltage source. As the loads change, they are placed in series with the Thevenin's equivalent resistance and voltage. The circuit is simply, a voltage source and two resistances in series. The voltage across the load and the current through the load may be determined using series circuit concepts. The Thevenin's equivalent circuit is determined in the following manner:

1. Place a voltmeter between the points where the load would be, and determine the voltage. This will be the Thevenin's voltage.
2. Remove the voltage source and replace it with a short.
3. Open any current sources in the circuit.
4. Place an ohmmeter between the points where the load would be, and determine the resistance seen from this point. This will be the Thevenin's resistance.
5. Redraw the equivalent circuit using the Thevenin's voltage and resistance as shown (Figure 19-1).

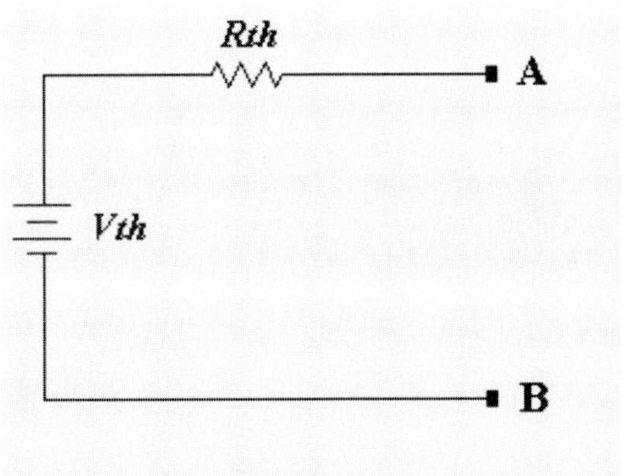

Figure 19-1

Learning Objectives

Upon completion of this lab, the student will be able to:

1. Given a series-parallel circuit, determine the Thevenin's equivalent circuit.
2. Using the Thevenin's equivalent circuit, determine the voltage across or current through various loads.

Pre-Lab Procedure

1. Using the procedure for determining Thevenin's equivalent circuit, predict Thevenin's voltage and resistance for the circuit in Figure 19-1. (Rl is the load resistor and V1=11.0V.) Record your results in Table 19-1.

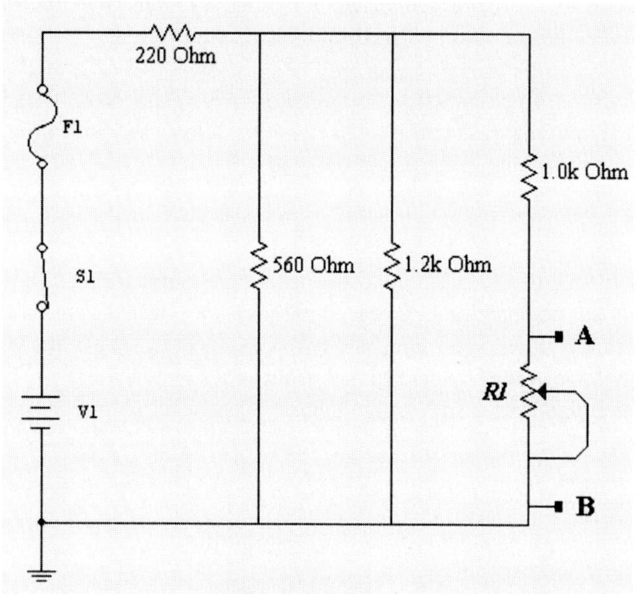

Figure 19-1

2. Using simulation software of your choice, predict voltage drops across and currents through the various load resistors. Use the circuit in Figure 19-1 to perform this analysis (V1=11.0V). Record your results in Table 19-2 (column – Simulated).

Note: *You are not asked to perform the analysis on the circuit in Figure 19-1 by hand. You are asked to simulate this circuit and predict various load resistor voltages and currents. Manual analysis on this circuit would be rather involved, especially if it had to be repeated for different load resistor values. This is an excellent time to mention the value of creating a Thevenin's equivalent circuit, as it eliminates the necessity for involved and repetitive circuit analysis. Thevenin's equivalent circuit reduces the complexity of predicting various load resistor voltages and currents.*

3. Using series circuit concepts, predict voltage drops across, and currents through, the various load resistors. Use the Thevenin's equivalent circuit in Figure 19-2 to perform this analysis. Record your results in Table 19-2 (column – Predicted).

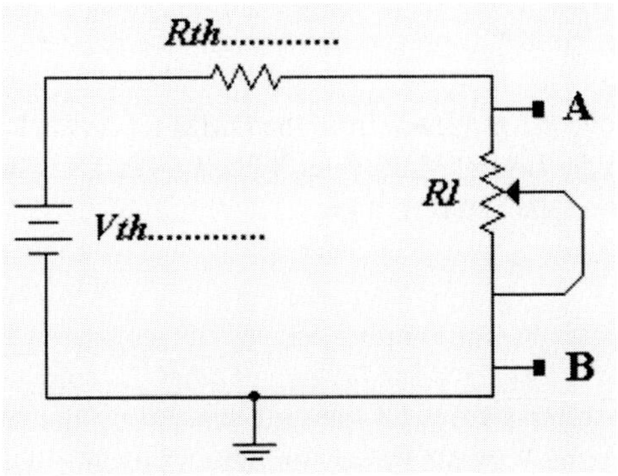

Figure 19-2
Thevenin's equivalent circuit

Lab Procedure

Before proceeding further, make sure that your eTCB is working correctly. Using eTCB Diagnostic VI, perform a self-diagnostic of eTCB circuitry.

Turn the switch S1 and S2 OFF in preparation for the lab.

Setup

Step #1 Using the eTCB Configuration VI, (DC module), reset all relays and then configure the circuit in Figure 19-3.

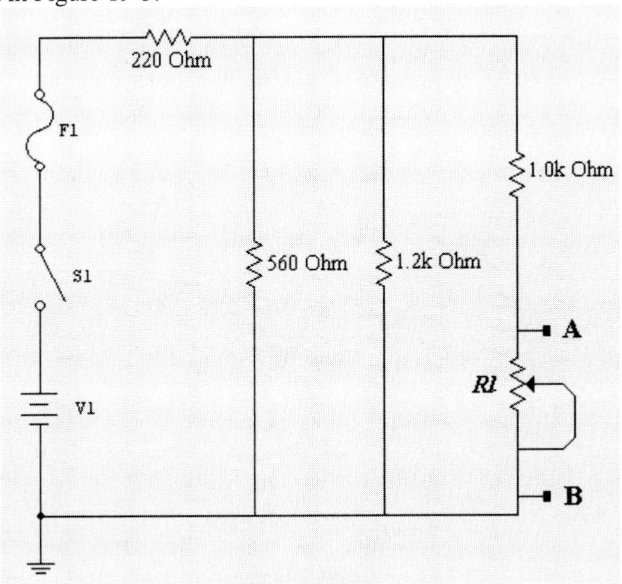

Figure 19-3

You will be asked to measure a DC current in this experiment, and as part of the DMM setup you need to "Null" the ammeter. This will minimize the error in DC current measurement. The "Nulling" procedure is very simple and needs to be performed only once per experiment. The procedure to "Null" the ammeter is as follows:

- Remove the test leads from the DMM terminals labeled "CURRENT".
- Open the Digital Multimeter VI and select DC Current measurement function.
- Click on the "Null" button.

Step #2 Using the Variable Power Supplies VI, set voltage source V1 to **11.0 V** . Keep in mind that this setting is just a good indication of what the voltage of V1 is set to. Variable power supplies often need to be adjusted (tuned) in order to set their output voltage precisely. Keep the Variable Power Supplies VI ***open***; if you close it, all the variable voltage supplies will reset to 0V.

Step #3 Using the Digital Multimeter VI, measure V1 voltage with respect to ground.

Step #4 Physically turn switch S1 on the eTCB board ON. This will complete your circuit. To verify that the switch is ON (closed) and that the fuse is not blown; use the Digital Multimeter VI to measure the voltage at TP1 with respect to ground.

Discovery

Step #5 You are going to perform the first step in the procedure leading to determining Thevenin's equivalent circuit. In this step you are asked to ***"Place a voltmeter between the points where the load would be and determine the voltage. This will be the Thevenin's equivalent voltage."***

You first need to remove the load resistor from the original circuit. To do that you need to physically remove jumper I4 from its terminals (Figure 19-4). This will effectively remove POT R15, our load resistor from the circuit. For your convenience, you may store the jumper in the Jumper Storage provided on the eTCB

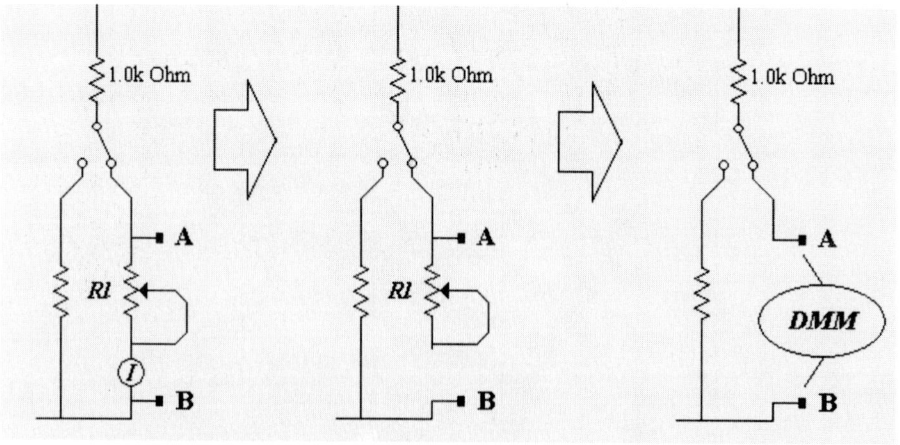

Figure 19-4
Removing jumper **I4** effectively removes load resistor from the circuit.

Using the Digital Multimeter VI, measure the voltage at **Point A** with respect to **Point B** (Figure 19-1). This is measured Thevenin's voltage. Record this reading in Table 19-1.

Step #6 You are now going to perform the second step in the procedure leading to determining Thevenin's equivalent circuit. In this step you are asked to ***"Remove the voltage source and replace it with a short."***

You need to reconfigure your circuit to remove V1 and replace it with a short circuit. Using the eTCB Configuration VI, reconfigure your circuit to remove V1 and replace it with a short.

Step #7 Since there are no current sources in this circuit, the third step in the procedure "***Open any current sources in the circuit***" does not need any attention. You are now going to perform the fourth step in the procedure leading to determining Thevenin's equivalent circuit. In this step you are asked to "***Place an ohmmeter between the points where the load would be, and determine the resistance seen from this point. This will be the Thevenin's resistance***".

With the jumper **I4** still removed, use the Digital Multimeter VI, to measure the resistance seen between Point A and Point B. This is measured Thevenin's resistance. Record this reading in Table 19-1.

Note: At this point, you have determined all the parameters of the Thevenin's equivalent circuit. The remaining portion of this lab will focus on validating the predicted load characteristics that you have determined in the pre-laboratory procedure.

Step #8 With the jumper I4 still removed, use the Digital Multimeter VI, to measure the resistance of POT R15 while adjusting R15 to read **300 Ω** (Figure 19-5).

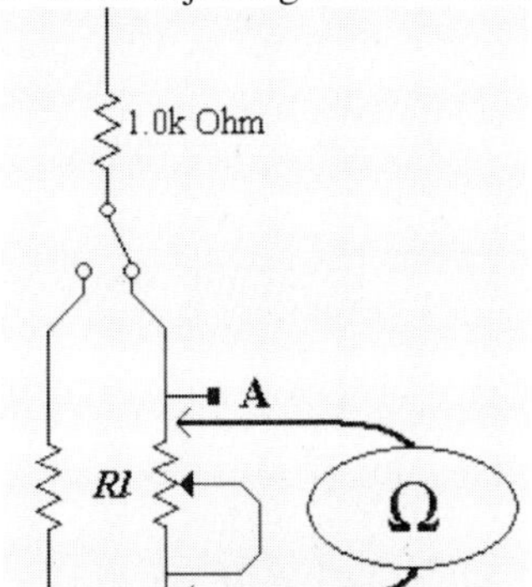

Figure 19-5

Step #9 With the jumper I4 still removed, use the eTCB Configuration VI, to reconfigure your circuit back to the original configuration. This configuration should configure the circuit in Figure 19-1 with load resistance of 300 Ω.

The reason why jumper I4 is still removed is because you are going to measure current through the load resistance next (Figure 19-6). Using the Digital Multimeter VI, measure the current through the load resistor. Record this reading in Table 19-2 (column – Measured).

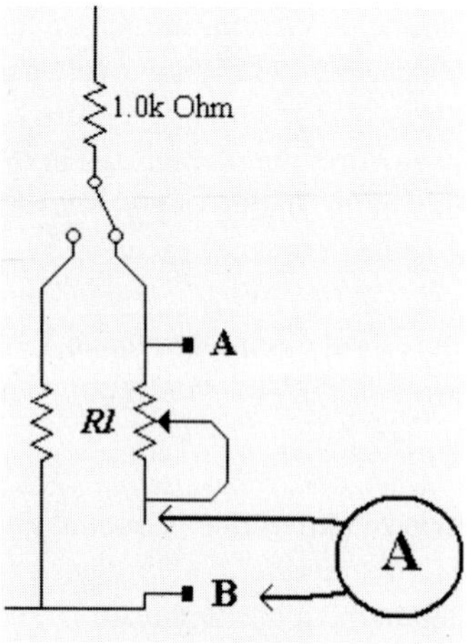

Figure 19-6

Step #10 Place jumper I4 in its original place. Using the Digital Multimeter VI, measure the voltage across the load resistor. Record this reading in Table 19-2 (column – Measured).

Step #11 You are now going to adjust load resistance to equal predicted Thevenin's resistance value. First you need to remove the jumper I4, and use the eTCB Configuration VI, to reconfigure your circuit to remove V1. This configuration removes V1 and allows for resistance measurement.

Now, use the Digital Multimeter VI to measure the resistance of POT R15 while adjusting R15 to equal the predicted Thevenin's resistance.

Step #12 With the jumper I4 still removed, use the eTCB Configuration VI, to reconfigure your circuit back to the original configuration. This configuration should configure circuit in Figure 19-1 with load resistance of R_{th}. The reason why jumper I4 is still removed is because you are going to measure current through the load resistance next.

Using the Digital Multimeter VI, measure the current through the load resistor. Record this reading in Table 19-2 (column – Measured).

Step #13 Place jumper I4 in its original place. Using the Digital Multimeter VI, measure the voltage across the load resistor. Record this reading in Table 19-2 (column – Measured).

Step #14 You are now going to adjust load resistance to **1.9k Ω.** First you need to remove the jumper I4, and use the eTCB Configuration VI, to reconfigure your circuit to remove V1. This configuration removes V1 and allows for resistance measurement. Now, use the Digital Multimeter VI to measure the resistance of POT R15 while adjusting R15 to equal 1.9k Ω.

Step #15 With the jumper I4 still removed, use the eTCB Configuration VI, to reconfigure your circuit back to the original configuration. This configuration should configure circuit in Figure 19-1 with load resistance of 1.9KΩ. The reason why jumper I4 is still removed is because you are going to measure current through the load resistance next.

Using the Digital Multimeter VI, measure the current through the load resistor. Record this reading in Table 19-2 (column – Measured).

Step #16 Place jumper I4 in its original place. Using the Digital Multimeter VI, measure the voltage across the load resistor. Record this reading in Table 19-2 (column – Measured).

This concludes the procedure for this experiment.

Table 19-1

Thevenin's Equivalent Circuits

	Predicted	***Measured***
Thevenin's Voltage		
Thevenin's Resistance		

Table 19-2

	Simulated	***Predicted***	***Measured***
Load Resistor Current			
$R_L = 300\ \Omega$			
$R_L = R_{th\ (Predicted)}$			
$R_L = 1.90k\ \Omega$			
Load Resistor Voltage			
$R_L = 300\ \Omega$			
$R_L = R_{th\ (Predicted)}$			
$R_L = 1.90k\ \Omega$			

Questions

1. What is the advantage of determining the Thevenin's equivalent circuit?

2. Perform the error calculation for all entries in Table 19-1 and Table 19-2. Comment on whether the errors are acceptable.

3. Use measured voltage and measured current through the various load resistances from Table 19-2 to determine the power delivered to the various load resistors.

	Delivered Power
$R_L = 300\ \Omega$	
$R_L = 1.14k\ \Omega$	
$R_L = 1.90k\ \Omega$	

4. Examine the results from the pervious question and determine the value of load resistance which caused the most power to be delivered. How does this value compare to the Thevenin's resistance?

5. To ensure the maximum power transfer to the load resistor, what must be the relationship between the Thevenin's resistance and load resistance?

Bonus:

6. When the value of load resistance is changed, how does it affect Thevenin's equivalent circuit?

20

Norton's Equivalent Circuit

Problem Statement

There often exists a situation where series-parallel circuits will have variable loads. Each time the load changes, the entire circuit would require complete analysis in order to determine the operating characteristics of the load. Consider a scenario where you do not know how the circuit is built: you only have access to its power source and the terminals where a load is connected. Your task is to determine the value of the load resistor and its dissipated power, at the condition of maximum power is transfer.

Norton's theorem analyzes the circuits from the point of view of the loads. The circuits are reduced to single current source with internal resistances. This approach allows an application of simplified circuit analysis to determine operating characteristics of the load.

Concept

In the case of a Norton's equivalent circuit, an equivalent resistance is placed in parallel with an equivalent current source. As the loads change, they are placed in parallel with the Norton's equivalent resistance and current source. The circuit is simply, a current source and two resistances in parallel. The voltage across the load and the current through the load may be determined using parallel-circuit concepts. The Norton's equivalent circuit is determined in the following manner:

1. Place an ammeter between the points where the load would be, and determine the current flow. This will be the Norton's current.
2. Remove the voltage source and replace it with a short.
3. Open any current sources in the circuit.
4. Place an ohmmeter between the points where the load would be, and determine the resistance seen from this point. This will be the Norton's resistance.
5. Redraw the equivalent circuit using the Norton's current and resistance as shown in Figure 20-1.

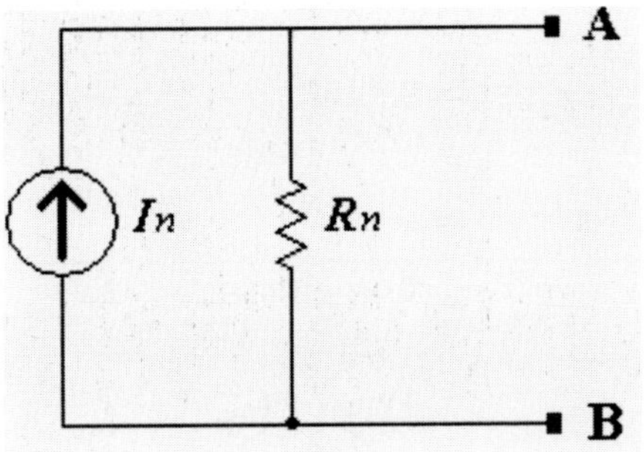

Figure 20-1

Learning Objectives

Upon completion of this lab, the student will be able to:

1. Given a series-parallel circuit, determine the Norton's equivalent circuit.
2. Using the Norton's equivalent circuit, determine the voltage across or current through various loads.

Pre-Lab Procedure

1. Using the procedure for determining Norton's equivalent circuit, predict Norton's voltage and resistance for the circuit in Figure 20-2. (Rl is the load resistor and V1=11.0V.) Record your results in Table 20-1.

2. Using simulation software of your choice; predict voltage drops across and currents through the various load resistors. Use the circuit in Figure 20-2 to perform this analysis (V1=11.0V). Record your results in Table 20-2 (column – Simulated).

Note: *You are not asked to perform the analysis on the circuit in Figure 20-2 by hand. You are asked to simulate this circuit and predict various load resistor voltages and currents. Manual analysis on this circuit would be rather involved, especially if it had to be repeated for different load resistor values. This is an excellent time to mention the value of creating a Norton's equivalent circuit, as it eliminates the necessity for involved and repetitive circuit analysis. Norton's equivalent circuit reduces the complexity of predicting various load resistor voltages and currents.*

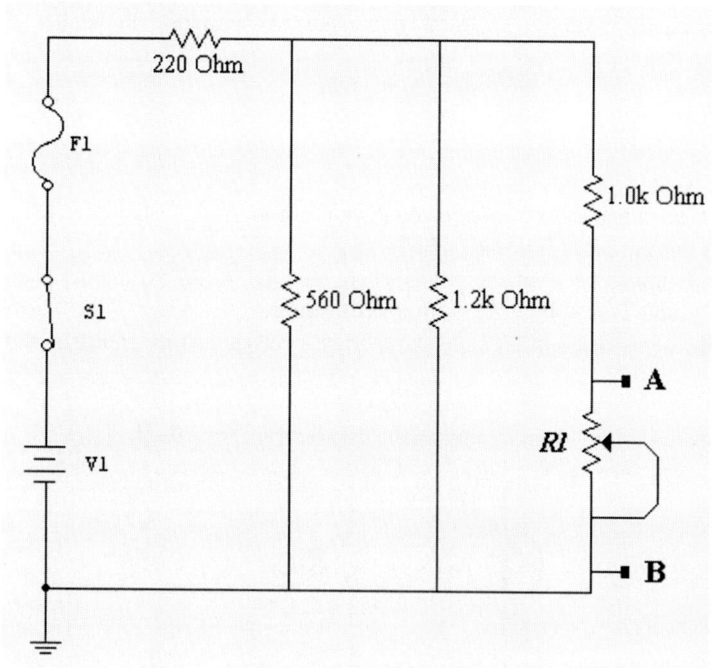

Figure 20-2

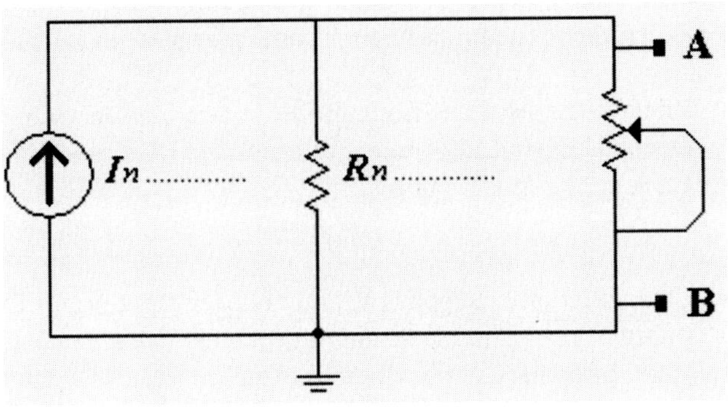

Figure 20-3

Lab Procedure

Setup

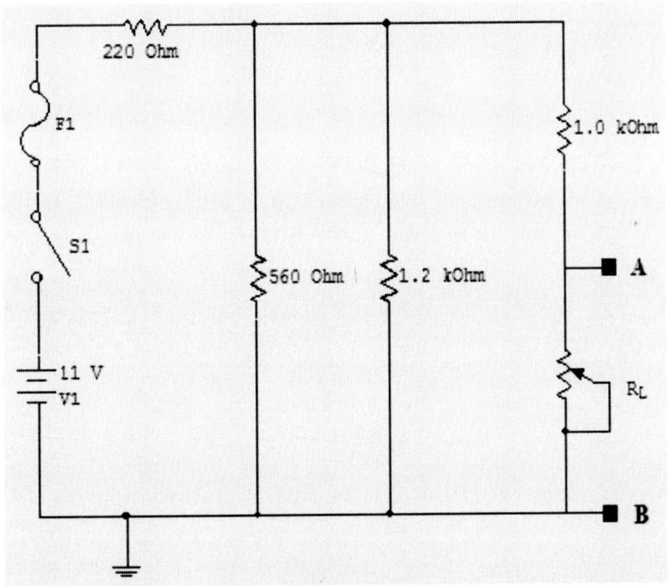

Figure 20-4

You will be asked to measure a DC current in this experiment, and as part of the DMM setup you need to "Null" the ammeter. This will minimize the error in DC current measurement. The "Nulling" procedure is very simple and needs to be performed only once per experiment. The procedure to "Null" the ammeter is as follows:

- Remove the test leads from the DMM terminals labeled "CURRENT"
- Open the Digital Multimeter VI and select DC Current measurement function.
- Click on the "Null" button.

Step #2 Using the Variable Power Supplies VI, set voltage source V1 to **11.0 V.** Keep in mind that this setting is just a good indication of what the voltage of V1 is set to. Variable power supplies often need to be adjusted (tuned) in order to set their output voltage precisely. Keep the Variable Power Supplies VI ***open***; if you close it, all the variable voltage supplies will reset to 0V.

Step #3 Using the Digital Multimeter VI, measure V1 voltage with respect to ground.

Step #4 Physically turn switch S1 on the eTCB board ON. This will complete your circuit. To verify that the switch is ON (closed) and that the fuse is not blown; use the Digital Multimeter VI to measure the voltage at TP1 with respect to ground.

Discovery

Step #5 You are going to perform the first step in the procedure leading to determining Norton's equivalent circuit. In this step you are asked to *"**Place an ammeter between the points where the load would be, and determine the current flow**"*. You first need to remove the load resistor from the original circuit. To do that, you need to physically remove jumper I4 from its terminals (Figure 20-5). This will effectively remove POT R15, our load resistor from the circuit. For your convenience you may store the jumper in the Jumper Storage provided on the eTCB board.

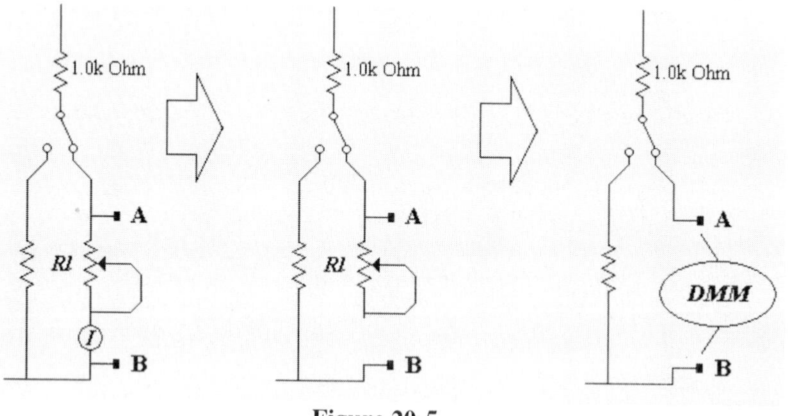

Figure 20-5

Removing jumper I4 effectively removes load resistor from the circuit.

Using the Digital Multimeter VI, measure the current flow between Point A and Point B (Figure 20-2). This is measured Norton's current. Record this reading in Table 20-1.

Step #6 You are now going to perform the second step in the procedure leading to determining Norton's equivalent circuit. In this step you are asked to "***Remove the voltage source and replace it with a short***". You need to reconfigure your circuit to remove V1 and replace it with a short circuit. Using the eTCB Configuration VI, reconfigure your circuit to remove V1 and replace it with a short.

Step #7 Since there are no current sources in this circuit the third step in the procedure "***Open any current sources in the circuit***" does not need any attention. You are now going to perform the fourth step in the procedure leading to determining Norton's equivalent circuit. In this step you are asked to "***Place an ohmmeter between the points where the load would be, and determine the resistance seen from this point. This will be the Norton's resistance***". With the jumper **I4** still removed, use the Digital Multimeter VI, to measure the resistance seen between Point A and Point B. This is measured Norton's resistance. Record this reading in Table 20-1.

Note: At this point you have determined all the parameters of the Norton's equivalent circuit. The remaining portion of this lab will focus on validating the predicted load characteristics that you have determined in the pre-laboratory procedure.

Step #8 With the jumper I4 still removed, use the Digital Multimeter VI, to measure the resistance of POT R15 while adjusting R15 to read **400 Ω** (Figure 20-6).

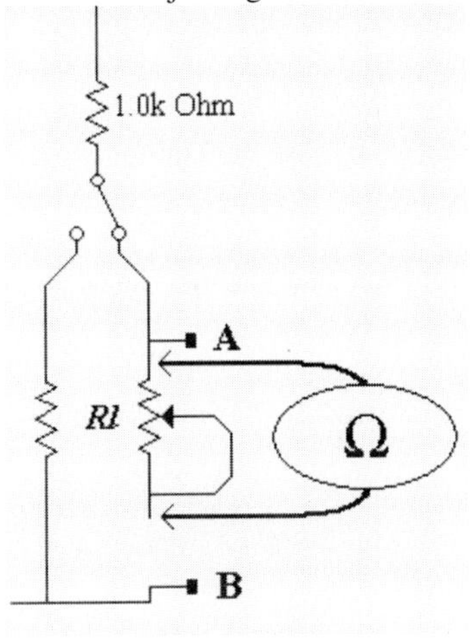

Figure 20-6

Step #9 With the jumper I4 still removed, use the eTCB Configuration VI, to reconfigure your circuit back to the original configuration. This configuration should configure circuit in Figure 20-2 with load resistance of 400 Ω. The reason why jumper I4 is still removed is because you are going to measure current through the load resistance next.

Using the Digital Multimeter VI, measure the current through the load resistor (Figure 20-7). Record this reading in Table 20-2 (column – Measured).

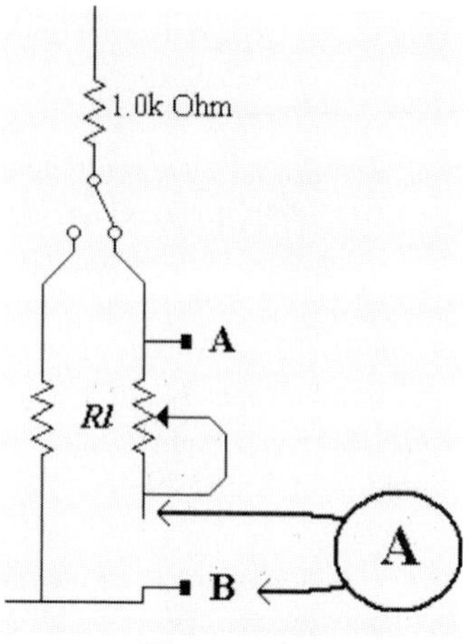

Figure 20-7

Step #10 Place jumper I4 in its original place. Using the Digital Multimeter VI, measure the voltage across the load resistor. Record this reading in Table 20-2 (column – Measured).

Step #11 You are now going to adjust load resistance to equal predicted Norton's resistance value. First you need to remove the jumper I4, and use the eTCB Configuration VI, to reconfigure your circuit to remove V1. This configuration removes V1 and allows for resistance measurement. Now, use the Digital Multimeter VI to measure the resistance of POT R15 while adjusting R15 to equal the predicted Norton's resistance.

Step #12 With the jumper I4 still removed, Use the eTCB Configuration VI, to reconfigure your circuit back to the original configuration. This configuration should configure circuit in Figure 20-2 with load resistance of $R_{N.}$ The reason why jumper I4 is still removed is because you are going to measure current through the load resistance next.

Using the Digital Multimeter VI, measure the current through the load resistor. Record this reading in Table 20-2 (column – Measured).

Step #13 Place jumper I4 in its original place. Using the Digital Multimeter VI, measure the voltage across the load resistor. Record this reading in Table 20-2 (column – Measured).

Step #14 You are now going to adjust load resistance to **1.8k Ω.** First you need to remove the jumper I4, and use the eTCB Configuration VI, to reconfigure your circuit to remove V1. This configuration removes V1 and allows for resistance measurement. Now, use the Digital Multimeter VI to measure the resistance of POT R15 while adjusting R15 to equal 1.8k Ω.

Step #15 With the jumper I4 still removed, use the eTCB Configuration VI, to reconfigure your circuit back to the original configuration. This configuration should configure circuit in Figure 20-2 with load resistance of 1.8k Ω. The reason why jumper I4 is still removed is because you are going to measure current through the load resistance next.

Using the Digital Multimeter VI, measure the current through the load resistor. Record this reading in Table 20-2 (column – Measured).

Step #16 Place jumper I4 in its original place. Using the Digital Multimeter VI, measure the voltage across the load resistor. Record this reading in Table 20-2 (column – Measured).

This concludes the procedure for this experiment.

Table 20-1
Norton's Equivalent Circuits

	Predicted	*Measured*
Norton's Current		
Norton's Resistance		

Table 20-2

	Simulated	*Predicted*	*Measured*
Load Resistor Current			
$R_L = 400\ \Omega$			
$R_L = R_{N\ (Predicted)}$			
$R_L = 1.80k\ \Omega$			
Load Resistor Voltage			
$R_L = 400\ \Omega$			
$R_L = R_{N\ (Predicted)}$			
$R_L = 1.80k\ \Omega$			

Questions

1. What is the advantage of determining the Norton's equivalent circuit?

2. Perform the error calculation for all entries in Table 20-1 and Table 20-2. Comment on whether the errors are acceptable.

3. Use measured voltage and measured current through the various load resistances from Table 20-2 to determine the power delivered to the various load resistors.

	Delivered Power
$R_L = 400\ \Omega$	
$R_L = 1.14k\ \Omega$	
$R_L = 1.80k\ \Omega$	

4. Examine the results from the previous question, and determine the value of load resistance that caused the most power to be delivered. How does this value compare to the Norton's resistance?

5. To ensure the maximum power transfer to the load resistor, what must be the relationship between the Norton's resistance and load resistance?

Bonus:

6. When the value of load resistance is changed, how does it affect Norton's equivalent circuit?

21

Superposition

Problem Statement

Consider having to solve a series-parallel circuit with multiple voltage sources. To address this challenge, the superposition method was developed. Adopting a concept of "dividing and conquering," this method allows for circuit analysis of circuits with one voltage source at a time.

Concept

Superposition analysis deals with the circuit, one voltage source at a time. The results are then superimposed, one on top of the other to determine the direction and amount of current flow. The process begins by removing one of the voltage sources and replacing it with a short and analyzing it as a circuit with one voltage source.

Learning Objectives

Upon completion of this lab, the student will be able to:

1. Configure a circuit using eTCB schematic drawing and eTCB board.
1. Determine the DC voltage across all components in a series-parallel circuit, using the superposition method.
1. Determine the individual currents using Ohm's law.

Pre-Lab Procedure

1. Given the voltage V1=9.0V, V2=-6.0V and V4=-15.0V, use the superposition method to predict the voltage drops across all resistive loads in circuit shown in Figure 21-1. Record your result in Table 21-1.

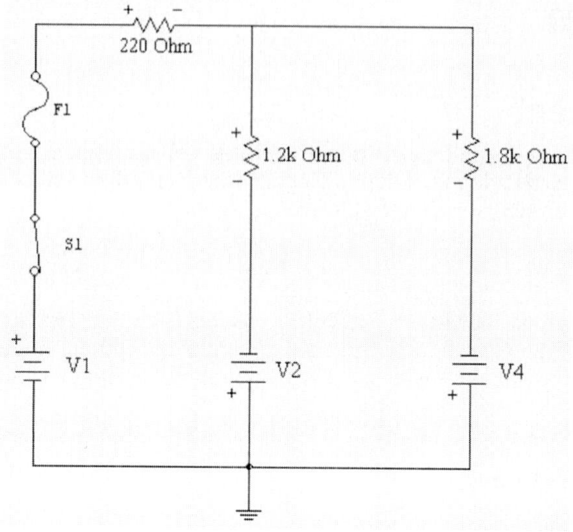

Figure 21-1

Lab Procedure

Before proceeding further, make sure that your eTCB is working correctly. Using eTCB Diagnostic VI, perform a self-diagnostic of eTCB circuitry.

Turn the switch S1 and S2 OFF in preparation for the lab.

Setup

Step #1 Using the Variable Power Supplies VI, set voltage source ***V1 to 9.0V*** and voltage source ***V2 to -6.0V.*** Keep in mind that this setting is just a good indication of what the voltage of V1 and V2 is set to. Variable power supplies often need to be adjusted (tuned) in order to set their output voltage precisely. Keep the Variable Power Supplies VI ***open***; if you close it, all the variable voltage supplies will reset to 0V.

Step #2 Using the Digital Multimeter VI, measure the output voltage of source **V1**. Place your DMM probes right at the voltage source terminals. Adjust the output of the variable power supplies, if necessary.

Step #3 Using the Digital Multimeter VI, measure the output voltage of source **V2**. Place your DMM probes right at the voltage source terminals. Adjust the output of the variable power supplies, if necessary.

Step #4 To verify the correct operation of voltage source **V4**, use the Digital Multimeter VI, to measure the output voltage of source V4. Place your DMM probes right at the voltage source terminals.

Discovery

*We will first measure the voltage drops due to voltage source **V1** alone.*

Step #5 Using the eTCB Configuration VI, (DC module), configure the circuit in Figure 21-2.

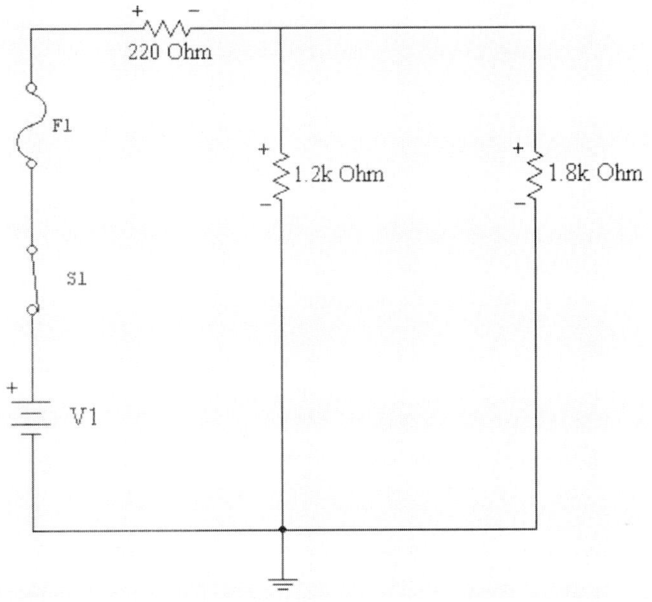

Figure 21-2

Step #6 Using the Digital Multimeter VI, measure voltage drop across **1.8k** _ resistor. Record this reading in Table 21-1 (column– Due to V1).

Step #7 Using the Digital Multimeter VI, measure voltage drop across **1.2k** _ resistor. Record this reading in Table 21-1 (column– Due to V1).

Step #8 Using the Digital Multimeter VI, measure voltage drop across **220** _ resistor. Record this reading in Table 21-1 (column– Due to V1).

*We will now measure the voltage drops due to voltage source **V2** alone.*

Step #9 Using the eTCB Configuration VI, (DC module), configure the circuit in Figure 21-3.

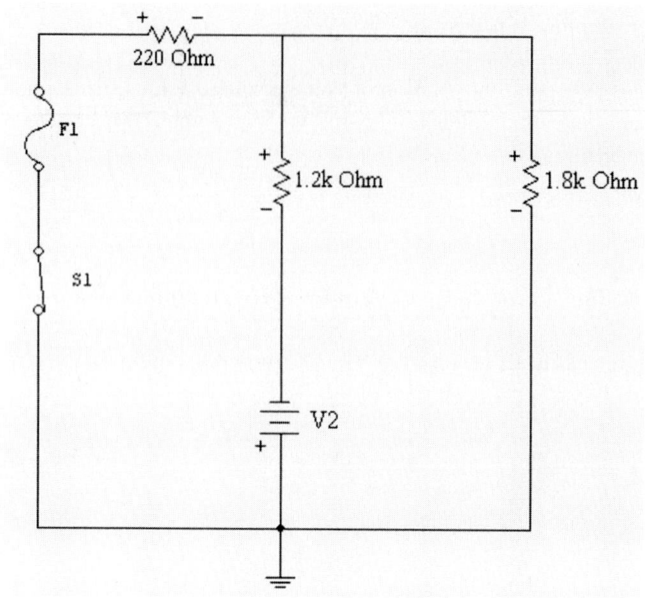

Figure 21-3

Step #10 Using the Digital Multimeter VI, measure voltage drop across **1.8k** _ resistor. Record this reading in Table 21-1 (column– Due to V2).

Step #11 Using the Digital Multimeter VI, measure voltage drop across **1.2k** _ resistor. Record this reading in Table 21-1 (column– Due to V2).

Step #12 Using the Digital Multimeter VI, measure voltage drop across **220** _ resistor. Record this reading in Table 21-1 (column– Due to V2).

*We will now measure the voltage drops due to voltage source **V4** alone.*

Step #13 Using the eTCB Configuration VI, (DC module), configure the circuit in Figure 21-4..)

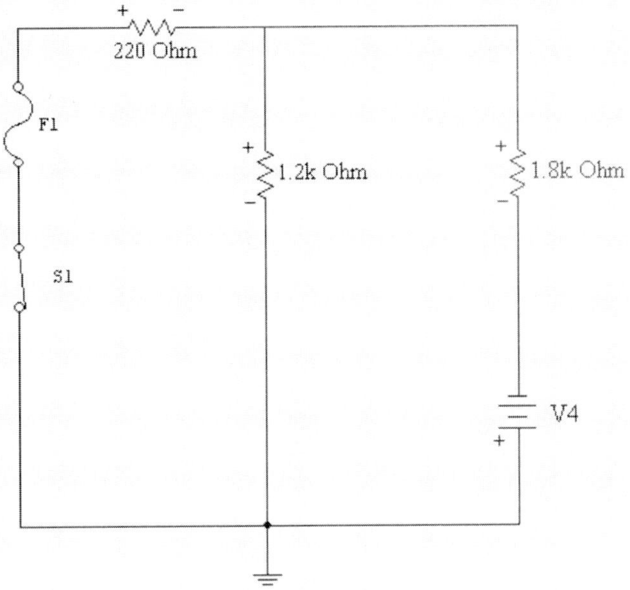

Figure 21-4

Step #14 Using the Digital Multimeter VI, measure voltage drop across **1.8k** _ resistor. Record this reading in Table 21-1 (column– Due to V4).

Step #15 Using the Digital Multimeter VI, measure voltage drop across **1.2k** _ resistor. Record this reading in Table 21-1 (column– Due to V4).

Step #16 Using the Digital Multimeter VI, measure voltage drop across **220** _ resistor. Record this reading in Table 21-1 (column– Due to V4).

We will finally measure the voltage drops due to all voltage sources.

Step #17 Using the eTCB Configuration VI, (DC module), configure the circuit in Figure 21-5.

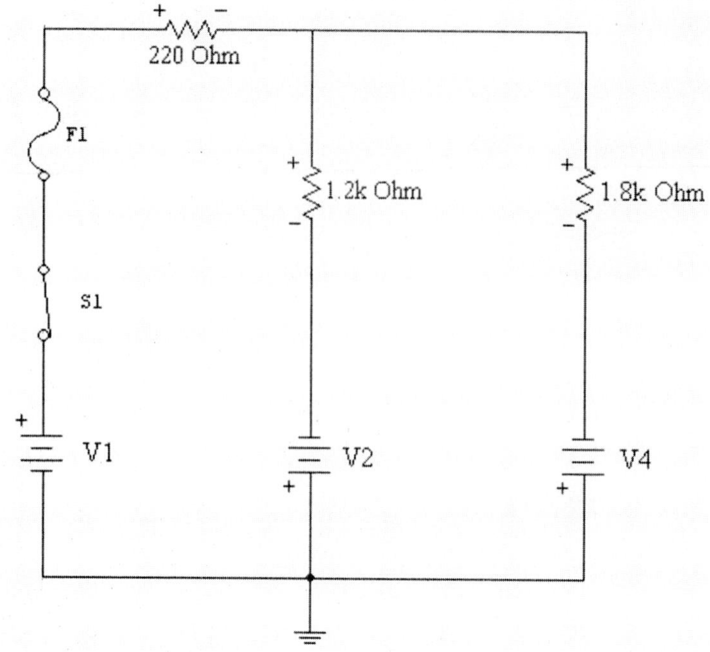

Figure 21-5

Step #18 Using the Digital Multimeter VI, measure voltage drop across **1.8k** _ resistor. Record this reading in Table 21-1 (column– Due to All).

Step #19 Using the Digital Multimeter VI, measure voltage drop across **1.2k** _ resistor. Record this reading in Table 21-1 (column– Due to All).

Step #20 Using the Digital Multimeter VI, measure voltage drop across **220** _ resistor. Record this reading in Table. (Column ***column– Due to All).***

This**cludes *the procedure for this experim*ent.**

Table 21-1

Predicted-Voltage Measured-Voltage Due to Voltage Source V1 *1.8k _ resistor 1.2k _ resistor 220 _ resistor* ***Due to Voltage Source V2***

1.8k _ resistor		
1.2k _ resistor		
220 _ resistor		

Due to Voltage Source V4		
1.8k _ resistor		
1.2k _ resistor		
220 _ resistor		
Due to All Voltage Sources together		
1.8k _ resistor		
1.2k _ resistor		
220 _ resistor		

Questions

1. What is the advantage of using the superposition method for solving complex circuits?

1. Perform the error calculation for all entries in Table 21-1, Comment on whether the errors are acceptable.

1. Use the Ohm's law to determine the currents through each resistor load in circuit shown in Figure 21-5

Bonus:

1. Can the superposition method be applied to solving circuits with both voltage and current sources?

22

Mesh Current

Concept

Series-parallel circuits with multiple voltage sources generally have multiple paths for current to flow. There are several methods of evaluating the circuits. One common method is the mesh current method. Using the mesh current method in order to evaluate a circuit such as the one in Figure 22-1, two current paths must be assumed. One of the loads in the circuit must possess the total current from the two assumed paths. These are randomly assumed paths. Once the current paths are determined, the polarities must be determined based on the assumed current paths. The polarities of the voltage sources are determined by the voltage sources.

Learning Objectives

Upon completion of this lab, the student will be able to:

1. Configure a circuit using eTCB schematic drawing and eTCB board.
1. Determine the DC current and direction of current through each load resistor using mesh current analysis.
1. Determine the individual voltage drops using Ohm's law.

Pre-Lab Procedure

1. Given the voltage V1=7.0V, and V4=15.0V, use mesh current analysis to predict currents I_A, I_B, I_C in the circuit shown in Figure 22-1. Use Ohm's law to predict voltage drops across all resistive loads. Pay attention to the indicated polarities. Record your result in Table 22-1.

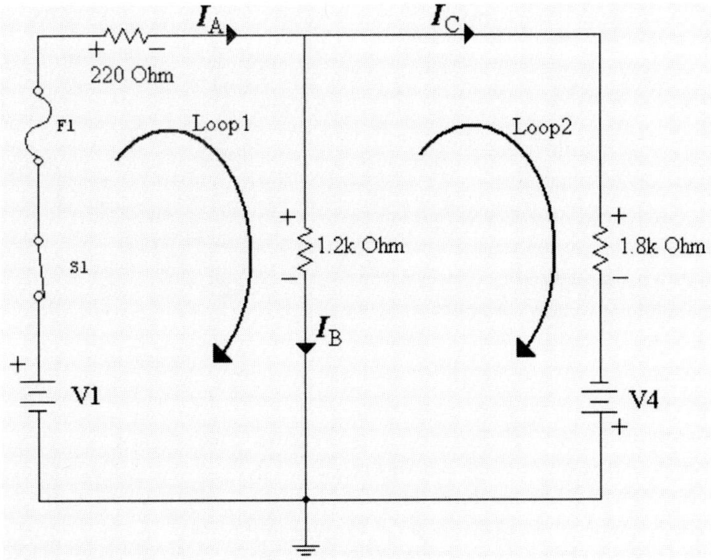

Figure 22-1

1. Repeat Pre-Lab Procedure 1 with the voltage V1=11.0V. Record your result in Table 22-1.

Lab Procedure

Before proceeding further, make sure that your eTCB is working correctly. Using eTCB Diagnostic VI, perform a self-diagnostic of eTCB circuitry.

Turn the switch S1 and S2 OFF in preparation for the lab.

Setup

Step #1 Using the eTCB Configuration VI, (DC module), reset all relays and then configure the circuit in Figure 22-2.

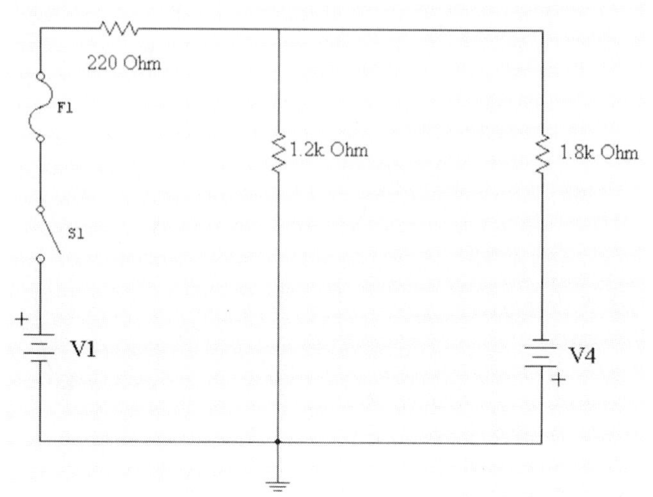

Figure 22-2

You will be asked to measure a DC current in this experiment, and as part of the DMM setup you need to "Null" the ammeter. This will minimize the error in DC current measurement. The "Nulling" procedure is very simple and needs to be performed only once per experiment. The procedure to "Null" the ammeter is as follows:

- Remove the test leads from the DMM terminals labeled "CURRENT".
- Open the Digital Multimeter VI and select DC Current measurement function.
- Click on the "Null" button.

Step #2 Using the Variable Power Supplies VI, set voltage source V1 to ***7.0V***. Keep in mind that this setting is just a good indication of what the voltage of V1 is set to. Variable power supplies often need to be adjusted (tuned) in order to set their output voltage precisely. Keep the Variable Power Supplies VI ***open***; if you close it, all the variable voltage supplies will reset to 0V.

Step #3 Physically turn switch S1 on the eTCB board "ON". To verify that the switch is ON (closed) and that the fuse is not blown, use the Digital Multimeter VI to measure voltage at **TP1** with respect to ground. Adjust the output of the variable power supplies, if necessary.

Step #4 To verify the correct operation of voltage source V4, use the Digital Multimeter VI, to measure voltage at **TP36** with respect to ground. This is the output voltage of voltage source V4.

Discovery

Step #5 Using the Digital Multimeter VI, measure the current through the **1.8k Ω** resistor. Record this reading in Table 22-1 (column – Measured). To perform this measurement, you need to insert the ammeter in series with the **1.8k Ω** resistor. Removing the jumper **I3** creates a break in the circuit where the ammeter can be inserted. Refer to Figure 22-3 for ammeter placement information. After the measurement is performed, make sure that every jumper is placed back in its place.

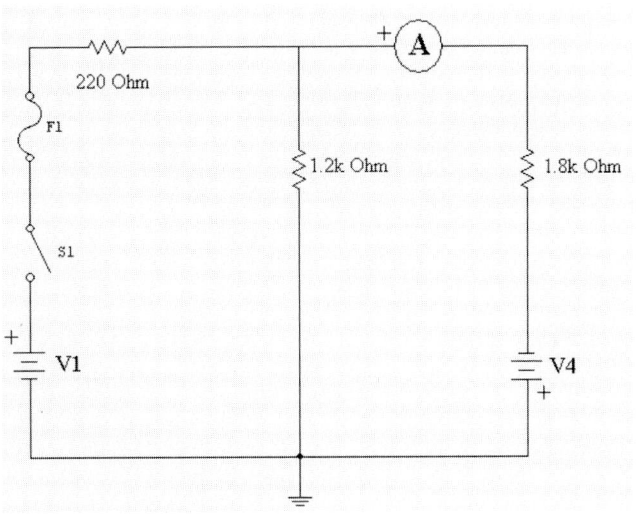

Figure 22-3

Step #6 Using the Digital Multimeter VI, measure the current through the **220 Ω** resistor. Record this reading in Table 22-1 (column – Measured). To perform this measurement, you need to insert the ammeter in series with the **220 Ω** resistor. Removing the jumper **I2** creates a break in the circuit where the ammeter can be inserted. Refer to Figure 22-4 for ammeter placement information. After the measurement is performed, make sure that every jumper is placed back in its place.

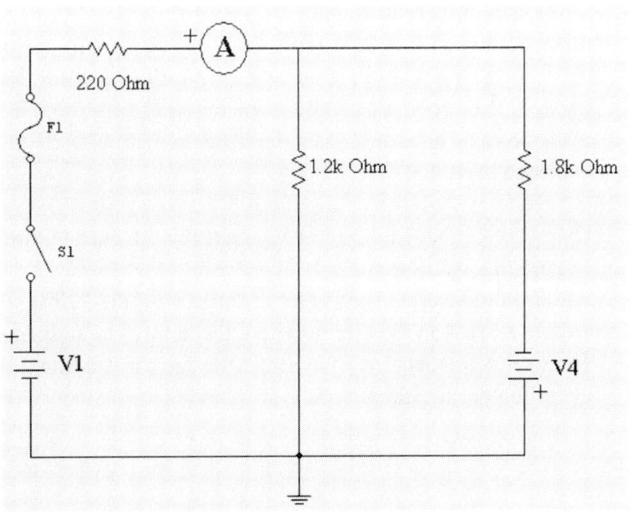

Figure 22-4

Step #7 Using the Digital Multimeter VI, measure the current through the **1.2k Ω** resistor. Record this reading in Table 22-1 (column – Measured). To perform this measurement, you need to insert the ammeter in series with the **1.2k Ω** resistor. Removing the jumper **J5** creates a break in the circuit where the ammeter can be inserted. Refer to Figure 22-5 for ammeter placement information. After the measurement is performed, make sure that every jumper is placed back in its place.

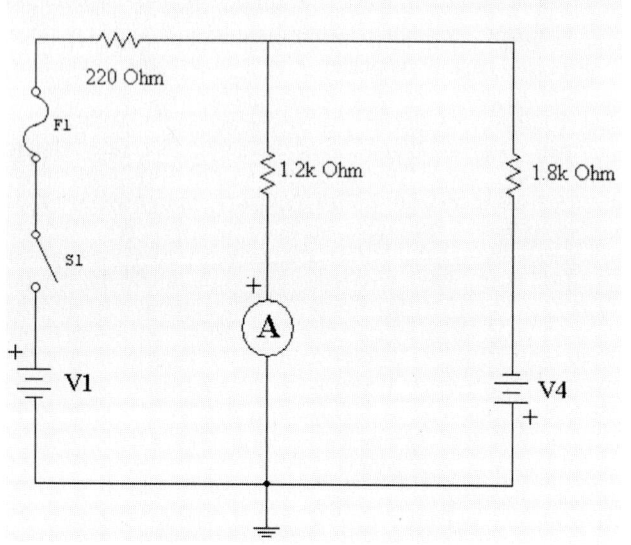

Figure 22-5

Note: We are now going to examine the effects of different supply voltage on the currents in the parallel circuit. We are going to use the same circuit diagram, but rather than setting V1=7.0V, we are going to set it at ***V1=11.0V.***

Step #8 Using the Variable Power Supplies VI, set voltage source V1 to ***11.0V***.
Using the Digital Multimeter VI, measure voltage at TP1 with respect to ground. Adjust the output of the variable power supplies, if necessary.

Step #9 Using the Digital Multimeter VI, measure the current through the **1.8k Ω** resistor. Record this reading in Table 22-1 (column – Measured). To perform this measurement, you need to insert the ammeter in series with the **1.8k Ω** resistor. Removing the jumper **I3** creates a break in the circuit where the ammeter can be inserted. Refer to Figure 22-3 for ammeter placement information. After the measurement is performed, make sure that every jumper is placed back in its place.

Step #10 Using the Digital Multimeter VI, measure the current through the **220 Ω** resistor..) Record this reading in Table 22-1 (column – Measured). To perform this measurement, you need to insert the ammeter in series with the **220 Ω** resistor. Removing the jumper **I2** creates a break in the circuit where the ammeter can be inserted. Refer to Figure 22-4 for ammeter placement information. After the measurement is performed, make sure that every jumper is placed back in its place.

Step #11 Using the Digital Multimeter VI, measure the current through the **1.2k Ω** resistor. Record this reading in Table 22-1 (column – Measured). To perform this measurement, you need to insert the ammeter in series with the **1.2k Ω** resistor. Removing the jumper **J5** creates a break in the circuit where the ammeter can be inserted. Refer to Figure 22-5 for ammeter placement information. After the measurement is performed, make sure that every jumper is placed back in its place.

This concludes the procedure for this experiment.

Table 22-1

	Predicted Current	**Measured Current**
V1=7.0V		
1.8k Ω		
220 Ω		
1.2k Ω		
V1=11.0V		
1.8k Ω		
220 Ω		
1.2k Ω		

Questions

1. What is the advantage of using the mesh current method for solving complex circuits?

1. Perform the error calculation for all entries in Table 22-1, Comment on whether the errors are acceptable.

1. Use the Ohm's law to determine voltage drops across each resistor load in circuit shown in Figure 22-1

Bonus:

1. Can the mesh current method be applied to solving circuits with both voltage and current sources?

23

Series-Parallel AC CircuitConcept

A resistive AC circuit is analyzed in the same manner as its DC counterpart. Recognizing that series-parallel circuits possess characteristics of both series and parallel circuits, voltage in these types of circuits will be determined accordingly. Voltage across parallel branches will be the same across all branches. Voltage developed across series loads is directly proportional to the resistance of the load and the current through the load.

The voltage across any load may be determined using Ohm's law. Once the voltage and resistance of the load are determined, they are divided to determine the current though the load. All series voltages are additive, all parallel voltages are the same.

Learning Objectives

Upon completion of this lab the student will be able to:

1. Configure a circuit using eTCB schematic drawing and eTCB board.
1. Given an AC supply voltage value, predict voltage drops on each resistor in a series-parallel AC circuit.
1. Measure the AC voltage in an AC series-parallel circuit.

Pre-Lab Procedure

1. Given the AC voltage V5=0.6Vrms, predict the voltages at point A and B with respect to ground in the series-parallel circuit shown in Figure 23-1. Record your result in Table 23-1.

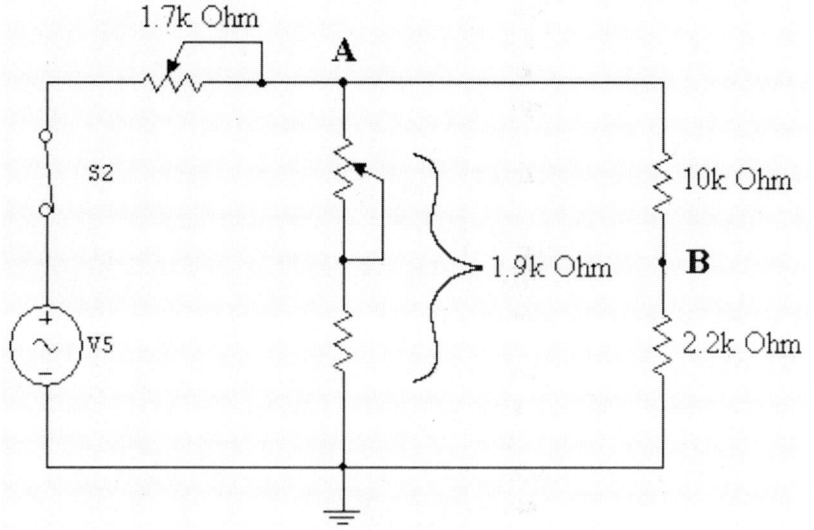

Figure 23-1

1. Repeat Pre-Lab Procedure 1 with the voltage V5=1.2Vrms. Record your result in Table 23-1.

Lab Procedure

Before proceeding further, make sure that your eTCB is working correctly. Using eTCB Diagnostic VI, perform a self-diagnostic of eTCB circuitry.

Turn the switch S1 and S2 OFF in preparation for the lab.

Setup

Step #1 Using the eTCB Configuration VI (AC module), reset all relays.

Step #2 With all relays still reset, use DMM VI, to measure resistance of POT R30. Adjust it to measure ***1.7k Ω.***.

Step #3 With all relays still reset, use DMM VI, to measure resistance of POT R34 in series with R35. Adjust their equivalent resistance to measure ***1.9k Ω.***.

Step #4 Using the eTCB Configuration VI, (AC Module) configure the circuit in Figure 23-2.

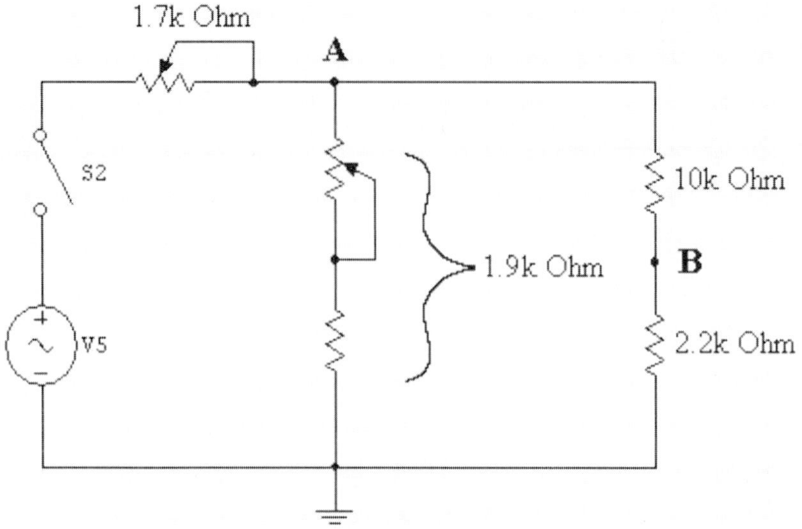

Figure 23-2

Step#4 (cont.) Using a Function Generator VI, set voltage source V5 to ***1.0V*** peak amplitude at ***500 Hz***. Keep in mind that this setting is just a good indication of what the voltage of V5 is set to. It may not necessarily be the actual voltage of V5. Amplitude of signals from function generators often need to be adjusted (tuned) in order to set their output voltage precisely. Keep the Function Generator VI ***open***; if you close it, its output will reset to 0V.

Step #5 Physically, turn the switch S2 on the eTCB board ON. This will complete your circuit. Using the Digital Multimeter VI, measure voltage at **TP40** with respect to ground. Adjust the amplitude of the output of the function generator to obtain a reading of ***0.6Vrms.***

Discovery

Step #6 Using the Digital Multimeter VI, measure the voltage at **point A** with respect to ground. Record this reading in Table 23-1 (column – Measured).

Step #7 Using the Digital Multimeter VI, measure the voltage at **point B** with respect to ground. Record this reading in Table 23-1 (column – Measured).

Note: We are now going to examine the effects of different supply voltage on the voltage drops across the individual resistors in the series-parallel AC circuit. We are going to use the same circuit diagram, but rather than setting V5=0.6Vrms, we are going to set it at ***V5=1.2Vrms.***

Step #8 Using a Function Generator VI, set voltage source V5 to ***2.0V*** peak amplitude at ***500 Hz***. Keep in mind that this setting is just a good indication of what the voltage of V5 is set to. It may not necessarily be the actual voltage of V5. Amplitude of signals from function generators often need to be adjusted (tuned) in order to set their output voltage precisely. Keep the Function Generator VI ***open***; if you close it, its output will reset to 0V.

Step #9 Using the Digital Multimeter VI, measure voltage at **TP40** with respect to ground. Adjust the amplitude of the output of the function generator to obtain a reading of ***1.2Vrms.***

Step #10 Using the Digital Multimeter VI, measure the voltage at **point A** with respect to ground. Record this reading in Table 23-1 (column – Measured).

Step #11 Using the Digital Multimeter VI, measure the voltage at **point B** with respect to ground. Record this reading in Table 23-1 (column – Measured).

This concludes the procedure for this experiment.

Table 23-1

	Predicted Voltage	*Measured Voltage*
V5=0.6Vrms		
V_A		
V_B		
V5=1.2Vrms		
V_A		
V_B		

Questions

1. Perform the error calculation for all entries in Table 23-1. Comment on whether the errors are acceptable.
1. Using predicted and measured data from Table 23-1, calculate voltages cross each resistive load in the circuit in Figure 23-1.
1. Using predicted and measured data from Table 23-1, calculate current through each resistive load in the circuit in Figure 23-1.
1. Using measured data from Table 23-1, demonstrate that Kirchhoff's voltage and current law applies to series-parallel AC circuits.
1. Using predicted and measured data from Table 23-1, calculate power dissipated in each resistive load in the circuit in Figure 23-1.

24

DC Voltage and Current in Capacitive Circuit

Problem Statement

Consider a car stereo system where a CD player is powered by the 12V battery. Being so close to a high voltage spikes caused by the engine electronic ignition system, in such a stereo system there are voltage spikes induced on the power lines supplying the power to the stereo system. That unwanted disturbance propagates through the stereo system and can be heard on the speakers. Your challenge is to build a circuit that would minimize the effects of the voltage spices on the quality of the sound from the stereo system. Understanding the behavior or capacitive circuits will help you in solving this challenge.

Concept

Capacitors resist a change in voltage. Capacitors develop a voltage as they are charged by the charging voltage. The voltage across a capacitor increases with time. After 5 RC time constants (__= R _ C), the capacitor is said to be at full charge. As the capacitor voltage increases, it practically reaches the charging voltage, at that point; there is no net difference of potential in the circuit. With no net voltage, there will be no current flow. The capacitor appears open to DC. For a brief instant in time, before the capacitor has developed a voltage, it appears as a short. In each time constant, the capacitor will develop approximately 63 percent of the difference of potential between the capacitor voltage and the forward charging voltage. Again, it takes five time constants for the capacitor to come to full charge.

$$Vc=Veexp(-t/T)$$

Learning Objectives

Upon completion of this lab the student will be able to:

1. Determine the DC voltage and current in a simple RC series circuit.
1. Determine the time constant for a simple RC series circuit.
1. Measure the DC voltage and current in a simple RC series circuit.

Pre-Lab Procedure

1. Predict the charge time constant and discharge time constant for the circuits in Figure 24-1 and Figure 24-2.

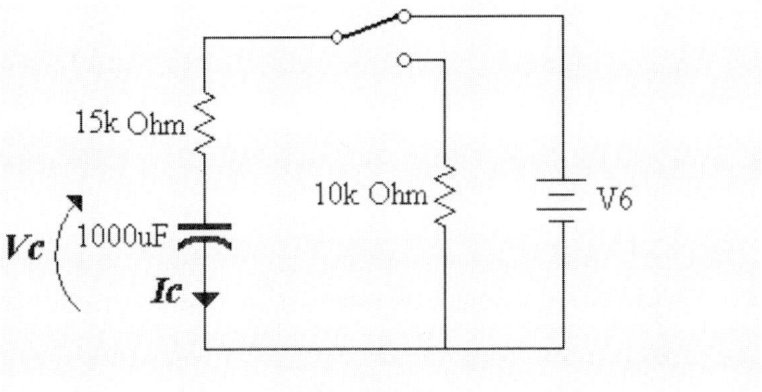

Figure 24-1

1. Given the charging voltage V6=15V and initial capacitor voltage is 0V, predict the capacitor voltage at different times as it is charged. Refer to the circuit in Figure 24-1. Record your result in Table 24-1.

1. Given the initial capacitor voltage is 15V, predict the capacitor voltage at different times as it is discharged. Refer to the circuit in Figure 24-2. Record your result in Table 24-2.

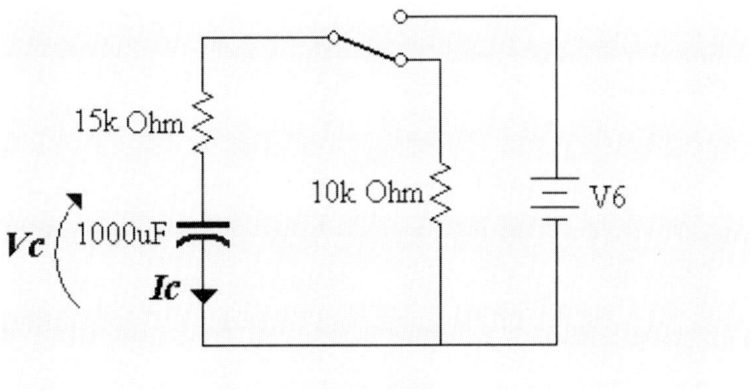

Figure 24-2

1. Given the charging voltage V6=15V and initial capacitor voltage is 0V, predict the capacitor current at different times as it is charged. Refer to the circuit in Figure 24-1. Record your result in Table 24-1.

1. Given the initial capacitor voltage is 15V, predict the capacitor current at different times as it is discharged. Refer to the circuit in Figure 24-2. Record your result in Table 24-2.

Lab Procedure

Before proceeding further, make sure that your eTCB is working correctly. Using eTCB Diagnostic VI, perform a self-diagnostic of eTCB circuitry.
Turn the switch S1 and S2 OFF in preparation for the lab.

Setup

Step #1 Using the eTCB Configuration VI, (AC module), reset all relays.

You will be asked to measure a DC current in this experiment, and as part of the DMM setup you need to "Null" the ammeter. This will minimize the error in DC current measurement. The "Nulling" procedure is very simple and needs to be performed only once per experiment. The procedure to "Null" the ammeter is as follows:

- Remove the test leads from the DMM terminals labeled "CURRENT".
- Open the Digital Multimeter VI and select DC Current measurement function.
- Click on the "Null" button.

Discovery

Note: *In this portion of the experiment we are going to gather voltage measurements to plot a capacitor voltage charging curve.*

Step #2 Using the Digital Multimeter VI, measure the capacitor voltage with respect to ground. This voltage should be close to 0V, indicating that the capacitor is discharged.

Step #3 Using the Digital Multimeter VI, measure the capacitor voltage with respect to ground, every 5 seconds, starting with the circuit configured to charge the capacitor. Record this reading in Table 24-1 (column – Measured). You can use ***"eTCB Timer"*** found in eTCB Extras, to help you with timing the measurements. It is best to use the "Single" button on the DMM to snap reading at the 5 seconds intervals.

Using the eTCB Configuration VI, (AC module), configure the capacitor charging circuit in Figure 24-3.

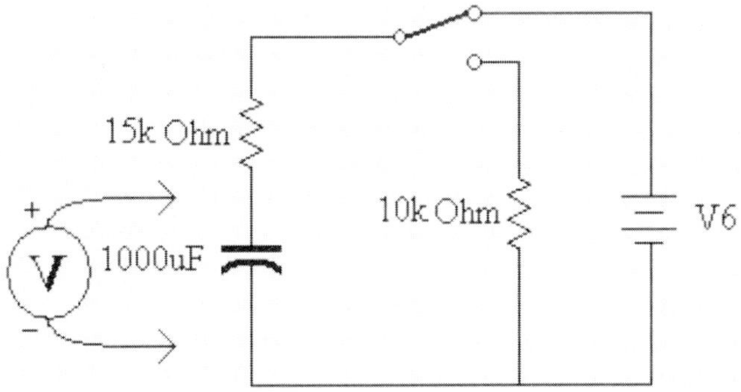

Figure 24-3

Note: *Make sure that the capacitor has been charging for at least 2 minutes. We want to make sure it is fully charged. We are now going to gather voltage measurements to plot a capacitor voltage-discharging curve.*

Step #4 Using the Digital Multimeter VI, measure the capacitor voltage with respect to ground. This voltage should be close to +15V, indicating that the capacitor is charged.

Step #5 Using the Digital Multimeter VI, measure the capacitor voltage with respect to ground, every 5 seconds, starting with the circuit configured to discharge the capacitor. Record this reading in Table 24-2 (column – Measured). You can use ***"eTCB Timer"*** found in eTCB Extras, to help you with timing the measurements. It is best to use the "Single" button on the DMM to snap reading at the 5 seconds intervals.

Using the eTCB Configuration VI, (AC module), configure the capacitor discharging circuit in Figure 24-4.

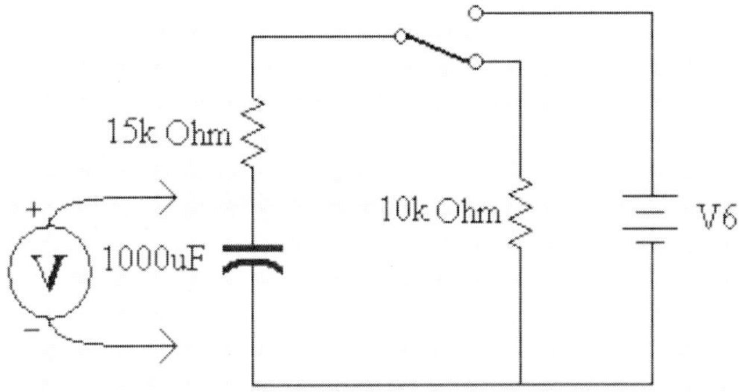

Figure 24-4

Note: *Make sure that the capacitor has been discharging for at least 2 minutes. We want to make sure it is fully discharged. We are now going to gather current measurements to plot a capacitor current-charging curve.*

Step #6 Using the Digital Multimeter VI, measure the current though a capacitor . This current should be close to 0A indicating that the capacitor is not charging or discharged. To perform this measurement, you need to insert the ammeter in series with the capacitor. Removing the jumper **I22** creates a break in the circuit where the ammeter can be inserted.

Step #7 Using the Digital Multimeter VI, measure the current though a capacitor, every 5 seconds, starting with the circuit configured to charge the capacitor. Record this reading in Table 24-1 (column – Measured). You can use ***"eTCB Timer"*** found in eTCB Extras, to help you with timing the measurements.

Using the eTCB Configuration VI, (AC module), configure the capacitor charging circuit in Figure 24-5.

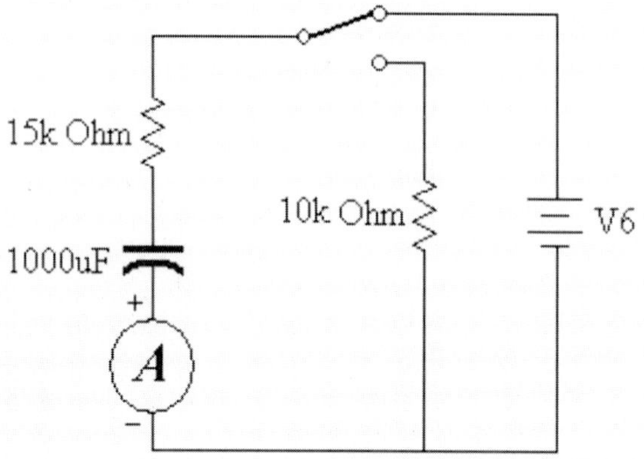

Figure 24-5

Note: *Make sure that the capacitor has been charging for at least 2 minutes. We want to make sure it is fully charged. We are now going to gather current measurements to plot a capacitor current-discharging curve.*

Step #8 Using the Digital Multimeter VI, measure the current though a capacitor . This current should be close to 0A indicating that the capacitor is not charging or discharged.

Step #9 Using the Digital Multimeter VI, measure the current though a capacitor, every 5 seconds, starting with the circuit configured to charge the capacitor. Record this reading in Table 24-2 (Column – Measured). You can use ***"eTCB Timer"*** found in eTCB Extras, to help you with timing the measurements.

Using the eTCB Configuration VI, (AC module), configure the capacitor discharging circuit in Figure 24-6.

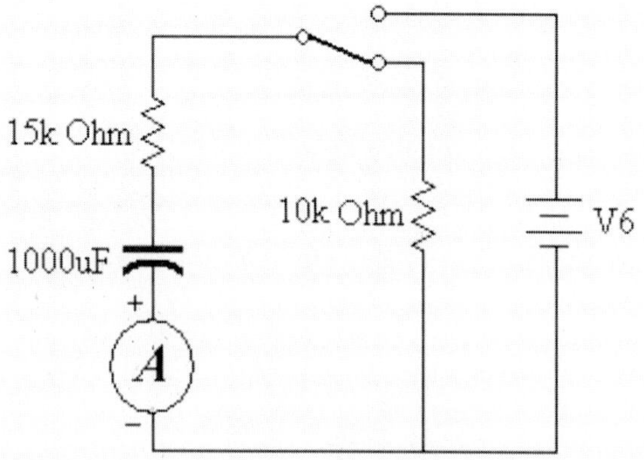

Figure 24-6

This concludes the procedure for this experiment.

Table 24-1 (Charging Capacitor)

Time	*Charging Capacitor Voltage*		*Charging Capacitor Current*	
	Predicted	***Measured***	***Predicted***	***Measured***
0 sec.				
5 sec.				
10 sec.				
15 sec.				
20 sec.				
25 sec.				
30 sec.				
35 sec.				
40 sec.				
45 sec.				
50 sec.				
55 sec.				
60 sec.				
65 sec.				
70 sec.				
75 sec.				
80 sec.				
85 sec.				
90 sec.				

Table 24-2 (Discharging Capacitor)

Time	*Discharging Capacitor Voltage*		*Discharging Capacitor Current*	
	Predicted	***Measured***	***Predicted***	***Measured***
0 sec.				
5 sec.				
10 sec.				
15 sec.				
20 sec.				
25 sec.				
30 sec.				
35 sec.				
40 sec.				
45 sec.				
50 sec.				
55 sec.				
60 sec.				
65 sec.				
70 sec.				
75 sec.				
80 sec.				
85 sec.				
90 sec.				

Questions

1. Perform the error calculation for all entries in Table 24-1. Comment on whether the errors are acceptable.

1. Using measured data from Table 24-1, determine the time constant for charging and discharging circuits. Comment on how it compares to the predicted ones.

1. Explain how the behavior exhibited in the capacitive circuits studied in this experiment can be used to solve the challenge posed in this experiment's Problem Statement.

Bonus:

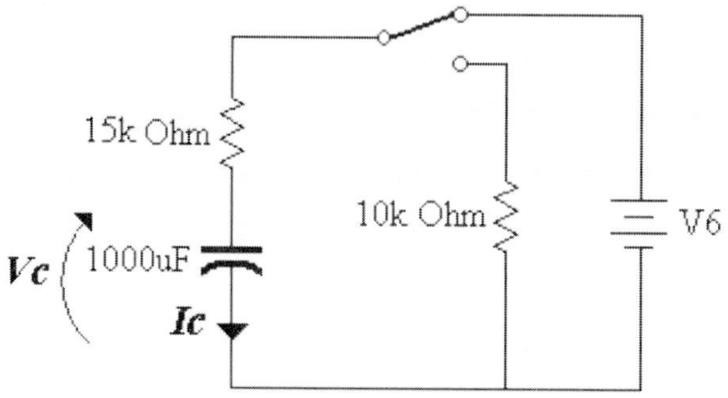

Figure 24-7

1. Given the charging voltage V6=20V and initial capacitor voltage is 5, predict the capacitor voltage two time constants later. Refer to Figure 24-7.

Troubleshooting

Exercise #1

Examining the charging and discharging capacitor voltage curves, you noticed that both the charge and discharge time constant were the same. Describe the scenario that would cause the circuit in Figure 24-8 to behave in the above manner.

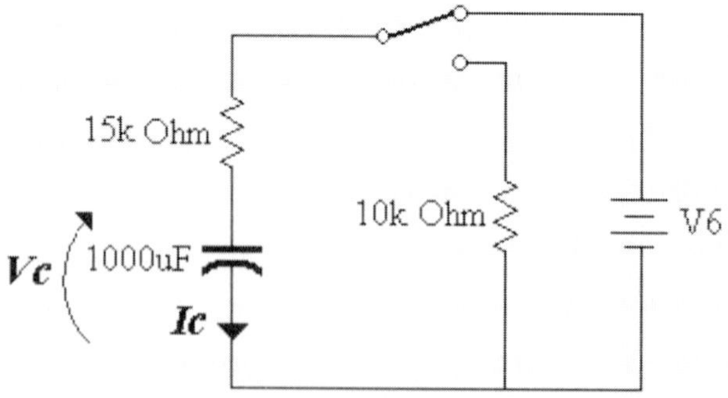

Figure 24-8

25

Using an Oscilloscope

Concept

An AC signal is different from a DC signal in that it changes direction and polarity. It is not enough simply to change magnitude; direction and polarity must be changed in order for an electrical signal to be an AC signal. AC signals are primarily sine waves, square waves, and triangle waves. These tend to be periodic waveforms—they repeat at anticipated times. AC signals vary in period and amplitude. An AC sine wave begins at point zero. It increases in amplitude until a maximum is reached, 90 into an alternation. If the increase was positive in direction, this will be the positive peak value. The signal will then begin to decrease in value until the value is zero. This occurs at the 180 point. The signal will then begin to increase in the opposite direction, changing the polarity of the signal until it reaches a maximum negative potential, 270 into the alternation. After the signal reaches a maximum negative voltage, the signal begins to increase until its amplitude is zero 360 into the alternation. At this point the signal repeats. This is referred to as one complete cycle of alternation.

When considering the value of an AC signal, there are three amplitude values to be familiar with. Measured from the positive peak to the negative peak is the peak-to-peak quantity. When measuring just the positive peak or the negative peak, this is known as the peak value (Figure 25-1). Peak value is half of the peak-to-peak value. Both peak-to-peak and peak values may be measured on the oscilloscope. Another value is the root-mean-square (RMS). RMS of the sine wave is calculated by multiplying the peak value by .707 or dividing by 1.414.

RMS value is also known as the effective value of the quantity.

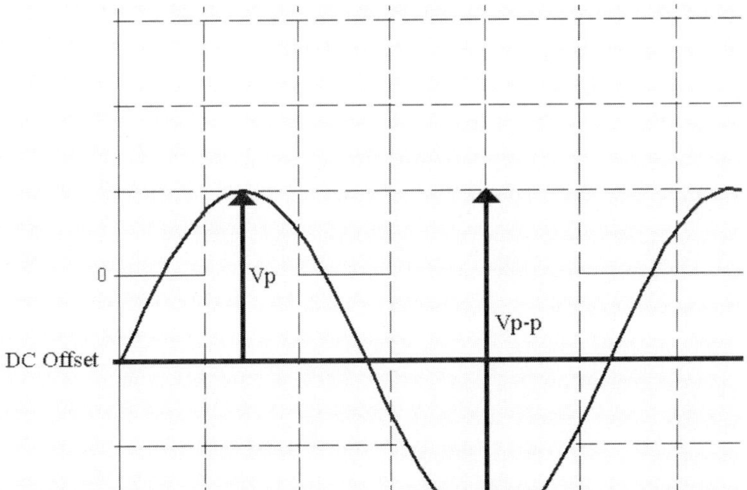

Figure 25-1

The amount of time necessary to complete one cycle of alternation is known as the period of the signal. The lower the period of the signal, the more often the signal repeats in one second. The greater the period, the less often the signal repeats in one second. The number of times a signal repeats in one second is known as the frequency of the signal. The period of the signal is the reciprocal of the frequency and is measured in seconds. The frequency of a signal is the reciprocal of the period and is measured in hertz, 1/period = frequency and 1/frequency = period. Period is represented in Figure 25-2. Frequency is determined the same way for sine waves, square waves, and triangle waves. Voltage, current, and power may also be represented as AC waveforms.

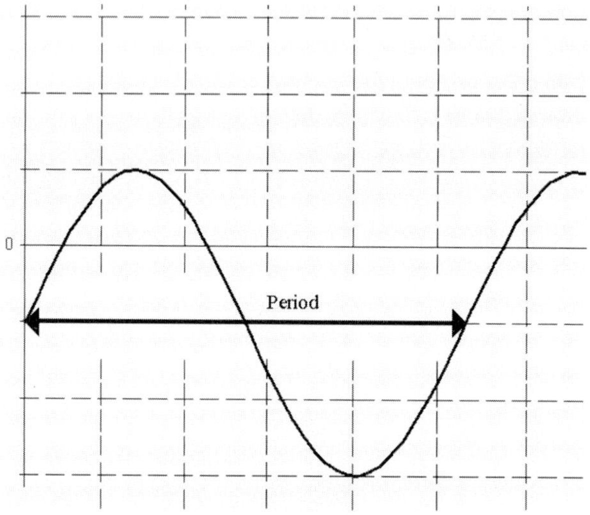

Figure 25-2 A

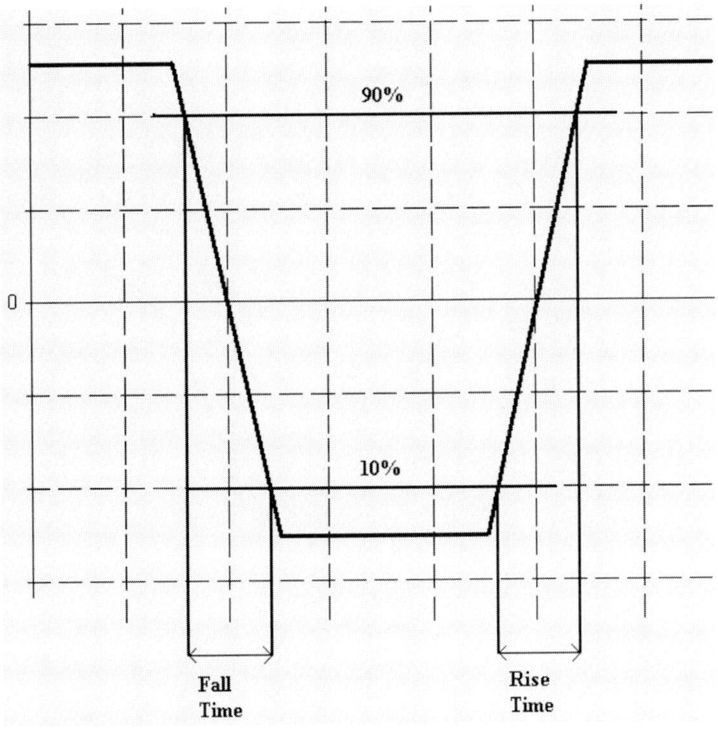

Figure 25-2 B

Learning Objectives

Upon completion of this lab, the student will be able to:

1. Use an oscilloscope to determine the peak, peak-to-peak, RMS and DC offset value of an AC signal.
1. Use an oscilloscope to determine the period, frequency, rise and fall time of an AC signal.
1. Use an oscilloscope to determine the phase angle between two AC signals.

Pre-Lab Procedure

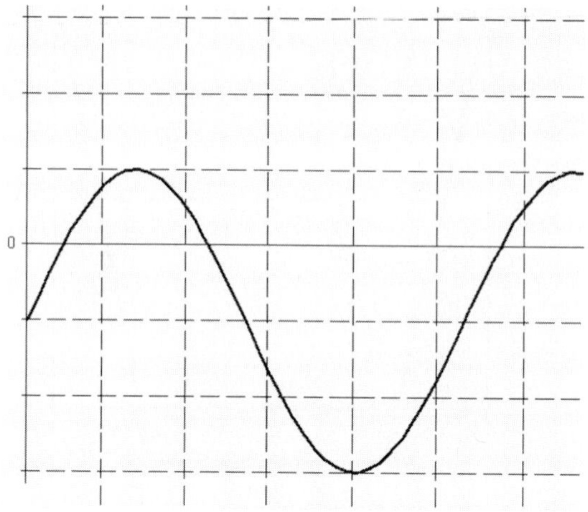

Figure 25-3

1. Consider an oscilloscope display shown in Figure 25-3. Determine peak, peak-to-peak, RMS and DC offset value of an AC signal, given the vertical scale of 500mV/div.

1. Consider an oscilloscope display shown in Figure 25-3. Determine the period and frequency of an AC signal, given the horizontal scale of 200_sec/div.

1. Consider an oscilloscope display shown in Figure 25-4. Determine the phase angle in degrees and radians between the two AC signals, given the horizontal scale of 200_sec/div.

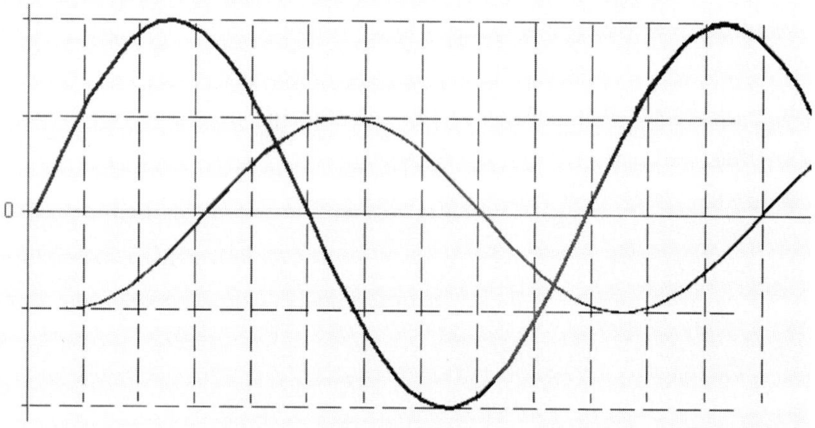

Figure 25-4

1. Consider an oscilloscope display shown in Figure 25-5. Determine both the rise time and the fall time of the AC signals, given the horizontal scale of 200_sec/div.

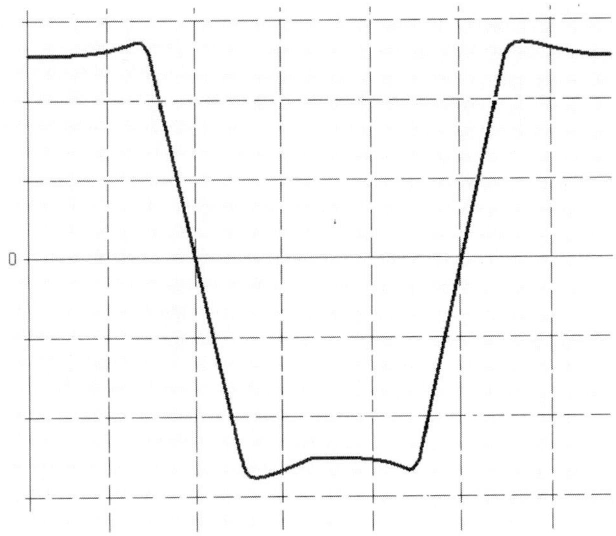

Figure 25-5

Lab Procedure

Before proceeding further, make sure that your eTCB is working correctly. Using eTCB Diagnostic VI, perform a self-diagnostic of eTCB circuitry.

Turn the switch S1 and S2 OFF in preparation for the lab.

Setup

Step #1 Using eTCB Configuration VI (AC module), reset all relays.

Step #2 With all relays still reset, use DMM VI, measure resistance of POT R34 in series with R35. Adjust its resistance to measure ***1.9k Ω***.

Step #3 With all relays still reset, use DMM VI, to measure the inductance of L1. Record it below:

Measured Inductance of L1	

Step #4 Using eTCB Configuration VI, (AC module) configure the circuit in Figure 25-6 The purpose to this circuit is to provide an output waveform that exhibits characteristics that we want to practice measuring. You will learn more about this circuit's characteristics in later experiments.

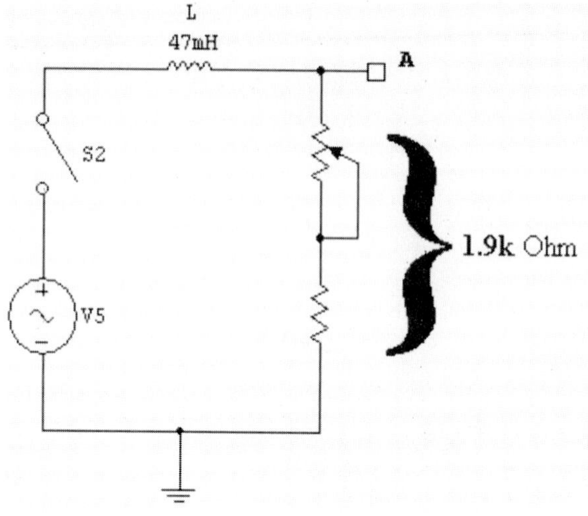

Figure 25-6

Step#4 (cont.) Using a Function Generator VI, set the voltage source V5 to a **sine waveform** with ***1.75V peak*** amplitude at ***100 Hz***. Keep in mind that this setting is just a good indication of what the voltage of V5 is set to. It may not necessarily be the actual voltage of V5. Amplitude of signals from function generators often need to be adjusted (tuned) in order to set their output voltage precisely. Keep the Function Generator VI ***open***; if you close it, its output will reset to 0V.

Step #5 Physically, turn the switch S2 on the eTCB board ON. This will complete your circuit.

Discovery

Step #6 Use Oscilloscope VI, to measure voltage at TP40 with respect to ground. Display this signal on CHANNEL A. Draw the waveform displayed on the oscilloscope.

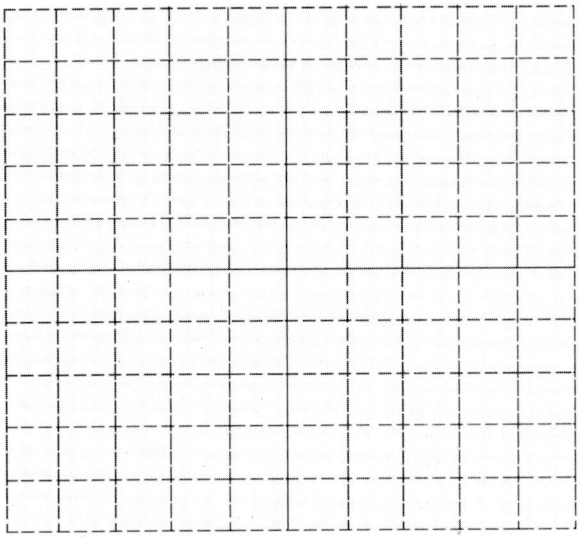

Channel A V/div....................

Channel B V/div....................

TimeBase sec/div....................

Figure 25-7

Step #7 Using a Function Generator VI, adjust the frequency of the V5 to ***600 Hz.*** Use Oscilloscope VI, to measure voltage at TP40 with respect to ground. Display this signal on CHANNEL A. Draw the waveform displayed on the oscilloscope.

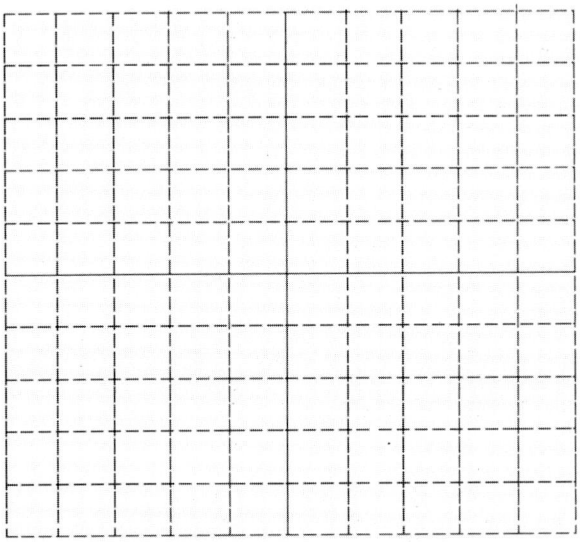

Channel A V/div....................

Channel B V/div....................

TimeBase sec/div....................

Figure 25-8

Step #8 Using a Function Generator VI, set the voltage source V5 to a **triangle waveform** with ***1.75V peak*** amplitude at ***2.0k Hz***. Keep in mind that this setting is just a good indication of what the voltage of V5 is set to. Amplitude of signals from function generators often need to be adjusted (tuned) in order to set their output voltage precisely.

Use Oscilloscope VI, to measure voltage at TP40 with respect to ground. Display this signal on CHANNEL A. Draw the waveform displayed on the oscilloscope.

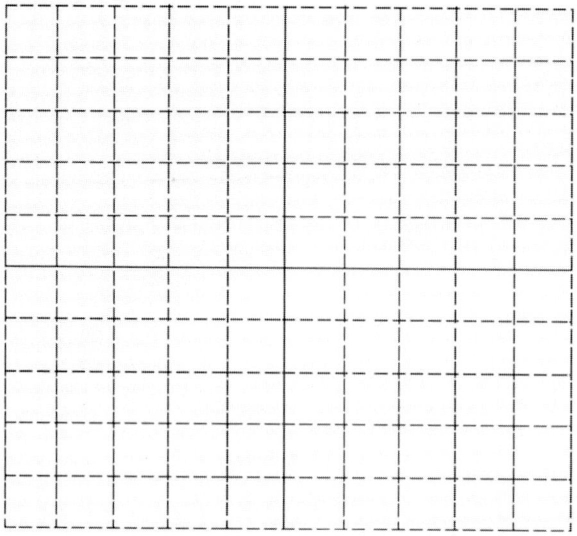

Channel A V/div....................

Channel B V/div....................

TimeBase sec/div...................

Figure 25-9

Step #9

Using the Function Generator VI, set the voltage source V5 to a **square waveform** with ***1.75V peak*** amplitude at ***4.0k Hz***. Keep in mind that this setting is just a good indication of what the voltage of V5 is set to. Amplitude of signals from function generators often need to be adjusted (tuned) in order to set their output voltage precisely.

Using the Oscilloscope VI, continue to measure the output of the function generator making sure that its output voltage remains at 3.5Vp-p. You may need to adjust it when it drifts too far away from 3.5Vp-p.

Use CHANNEL B of the oscilloscope to measure the voltage at point A with respect to ground (Figure 25-6).

Using cursors, measure the fall time of the signal displayed on channel B

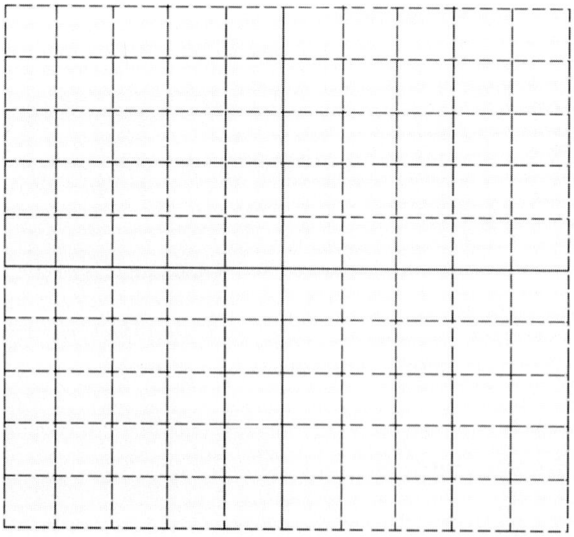

Channel A V/div....................

Channel B V/div....................

TimeBase sec/div...................

Figure 25-10

Step #10 Using the Function Generator VI, set the voltage source V5 to a **sine waveform** with ***1.75V peak*** amplitude at ***4.0k Hz***.

Using the Oscilloscope VI, continue to measure the output of the function generator making sure that its output voltage remains at 3.5Vp-p. You may need to adjust it when it drifts too far away from 3.5Vp-p.

Use CHANNEL B of the oscilloscope to measure the voltage at point A with respect to ground (Figure 25-6).

Using cursors measure the Phase Angle between the two signals. (Channel A and Channel B)

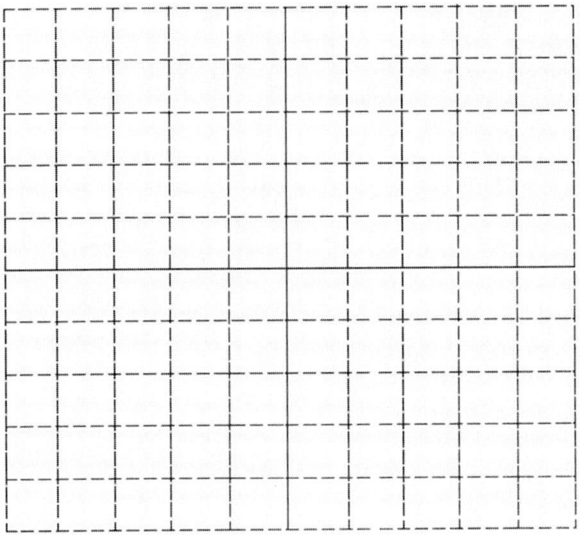

Channel A V/div....................

Channel B V/div....................

TimeBase sec/div....................

Figure 25-11

This concludes the procedure for this experiment.

Questions

1. Consider figure 25-7:
   What is the peak-to-peak value of the displayed waveform?
   What is the RMS value of the displayed waveform?
   What is the frequency of the displayed waveform?
   What is the period of the displayed waveform?

1. Consider figure 25-8:
   What is the peak-to-peak value of the displayed waveform?
   What is the RMS value of the displayed waveform?
   What is the frequency of the displayed waveform?
   What is the period of the displayed waveform?

1. Consider figure 25-9:
   What is the peak-to-peak value of the displayed waveform?
   What is the period of the displayed waveform?
   What is the rise time of the displayed waveform?
   What is the fall time of the displayed waveform?

1. Consider figure 25-10:
   What is the peak-to-peak value of the displayed waveforms on Channel A and B?
   What is the period of the displayed waveforms on Channel A and B?
   What is the rise time of the displayed waveform on Channel B?
   What is the fall time of the displayed waveform on Channel B?

1. Consider figure 25-11:
   What is the phase angle between the two waveforms?

26

RL Circuits

Problem Statement

Consider a stereo system with a subwoofer. The subwoofer is designed to produce the low frequency sounds, and any high frequency sounds would be distorted by the subwoofer. Your challenge is to take the full frequency range audio signal and build a circuit that would be placed before the subwoofer. The purpose of this circuit is to condition the input signal to the subwoofer, so that only the low frequency signals are present. Understanding the behavior of RL circuits for different input signals frequency will help you in solving your challenge.

Concept

Inductors possess a characteristic known as inductive reactance (XL). Inductive reactance is the opposition to current flow caused by the inductor and is measured in ohms. Inductive reactance is calculated using the formula $2\pi f L$, where f is frequency in Hertz and L is inductance in Henrys. It can be seen that inductive reactance is directly related to the frequency of the AC signal.

Inductive reactance is an imaginary value. On an X-Y coordinate system X_L is on the positive Y-axis, giving it an angle of +90°. The angle of the reactance with the resistance, which is at 0°, will produce a phase shift in current. If X_L is equal to resistance in a series RL circuit the total impedance (Zt) will have an angle of 45, as shown in Figure 26-1. This will cause the current to have an angle of -45°. Current will lag the voltage in an inductive circuit by 90°.

If X_L is less than the resistance in a series RL circuit, the angle of the impedance will be from 0° to 45° and current will have an angle from 0° to -45°. If X_L is greater than the resistance in a series RL circuit, the angle of the impedance will be from 45 to 90 and current will have an angle from -45° to -90°. Current and voltage are calculated using Ohm's law. As with all series circuits the greater the impedance in the circuit, the less the current in the circuit. In series circuits the circuit becomes more inductive at higher frequencies

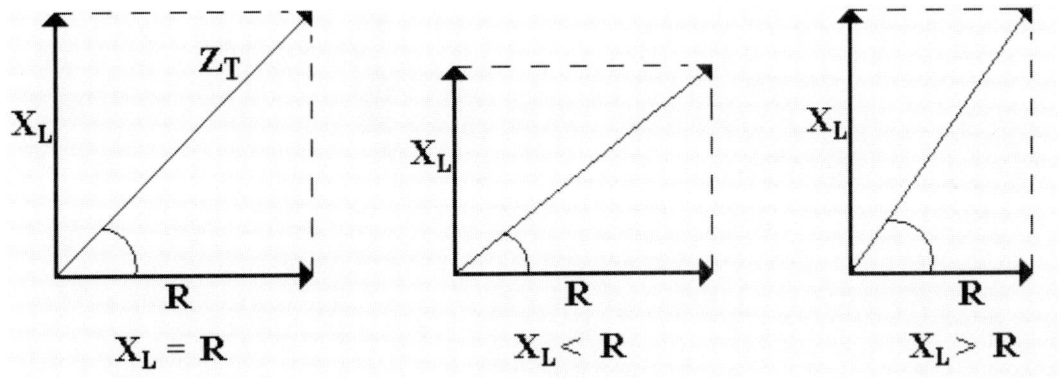

Figure 26-1

Learning Objectives

Upon completion of this lab, the student will be able to:

1. Configure an RL circuit using eTCB.
1. Determine inductive reactance at various frequencies.
1. Determine total impedance of an RL circuit.
1. Draw the phasor representation of the impedance of an RL circuit.
1. Determine the voltage and current in an RL circuit to include phase angles.
1. Measure the voltage in an RL circuit including phase angles.
1. Draw the phasor representation of the relationship of voltage and current in an RL circuit.

Pre-Lab Procedure

1. Predict the inductive reactance and total impedance of the circuit shown in Figure 26-2. Record your results in Table 26-1.

1. Draw a phasor diagram of resistance, inductive reactance and total impedance of the circuit shown in Figure 26-2. Use frequency values from Table 26-1

Table 26-1

Frequency	*Inductive Reactance*	*Total Impedance*
100 Hz		
600 Hz		
2.0k Hz		
10.0k Hz		

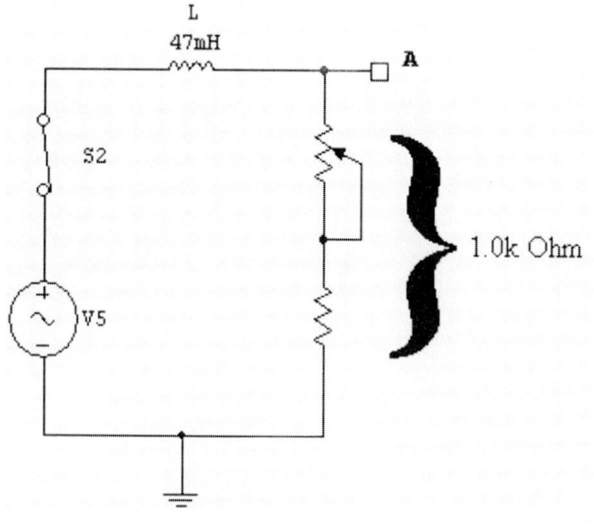

Figure 26-2

1. Predict the voltage (magnitude and phase) at point A of the circuit shown in Figure 26-2. Record your results in Table 26-2. V5=3.5Vp-p.

1. Repeat Pre-Lab procedure #3 for inductance value of 100mH. Record your results in Table 26-2.

Lab Procedure

Before proceeding further, make sure that your eTCB is working correctly. Using eTCB Diagnostic VI, perform a self-diagnostic of eTCB circuitry.

Turn the switch S1 and S2 OFF in preparation for the lab.

Setup

Step #1 Using eTCB Configuration VI (AC module), reset all relays.

Using DMM VI, measure the equivalent resistance of POT R34 and R35. Adjust its resistance to measure **1.0k Ω.**

Step #2 With all relays still reset, use DMM VI, to measure the inductance of L1. Record it below:

Measured Inductance of L1	

Step #3 With all relays still reset, use DMM VI, to measure the inductance of L2. Place jumper **J11** in the B to C position to select inductor L2. Record its inductance below:

Measured Inductance of L2	

Step #4 Using eTCB Configuration VI, (AC module) configure the circuit in Figure 26-3. Place jumper **J11** back in the A to C position to select inductor L1.

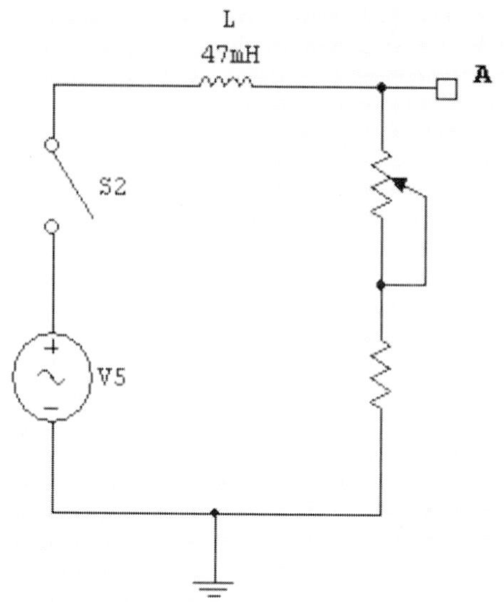

Figure 26-3

Step#4 (cont.) Using a Function Generator VI, set the voltage source V5 to ***1.75V peak*** amplitude at ***100 Hz***. Keep in mind that this setting is just a good indication of what the voltage of V5 is set to. It may not necessarily be the actual voltage of V5. Amplitude of signals from function generators often need to be adjusted (tuned) in order to set their output voltage precisely. Keep the function generator VI ***open***; if you close it, its output will reset to 0V.

Step #5 Physically, turn the switch S2 on the eTCB board ON. This will complete your circuit. Use Oscilloscope VI, to measure voltage at TP40 with respect to ground. Display this signal on CHANNEL A. Use the measurement option on the oscilloscope to measure signal parameters on CHANNEL A. Continue to perform this measurement while adjusting the signal amplitude with function generator. Adjust V5 to ***3.5Vp-p.***

Discovery

Step #6 Using an Oscilloscope VI, continue to measure the output of the function generator making sure that its output voltage remains at 3.5Vp-p. You may need to adjust it when it drifts too far away from 3.5Vp-p.

Use CHANNEL B of the oscilloscope to measure the voltage (magnitude and hase) at point A with respect to ground (Figure 26-3). Record this reading in Table 26-2.

Step #7 Using a Function Generator VI, set the frequency of the V5 to ***600 Hz.***
Using an Oscilloscope VI, continue to measure the output of the function generator making sure that its output voltage remains at 3.5Vp-p. Use CHANNEL B of the oscilloscope to measure the voltage (magnitude and phase) at point A with respect to ground (Figure 26-3). Record this reading in Table 26-2.

Step #8 Using a Function Generator VI, set the frequency of the V5 to ***2.0 kHz.***
Using an Oscilloscope VI, continue to measure the output of the function generator making sure that its output voltage remains at 3.5Vp-p. Use CHANNEL B of the oscilloscope to measure the voltage (magnitude and phase) at point A with respect to ground (Figure 26-3). Record this reading in Table 26-2.

Step #9 Using a Function Generator VI, set the frequency of the V5 to ***10.0 kHz.***
Using an Oscilloscope VI, continue to measure the output of the function generator making sure that its output voltage remains at 3.5Vp-p. Use CHANNEL B of the oscilloscope to measure the voltage (magnitude and phase) at point A with respect to ground (Figure 26-3). Record this reading in Table 26-2.

Turn the switch S2 "OFF".

Note: We are now going to examine the effects of different inductance value on the circuit behavior. We are going to use the same circuit diagram, but rather than choosing to use inductor L1, you are going to choose ***inductor L2*** *instead. Placing jumper J11 in the B to C position will select inductor L2.*

Turn the switch S2 "ON".

Step #10 Using a Function Generator VI, set the frequency of the V5 to ***100 Hz.***
Using an Oscilloscope VI, continue to measure the output of the function generator making sure that its output voltage remains at 3.5Vp-p. Use CHANNEL B of the oscilloscope to measure the voltage (magnitude and phase) at point A with respect to ground (Figure 26-3). Record this reading in Table 26-2.

Step #11 Using a Function Generator VI, set the frequency of the V5 to ***600 Hz.***
Using an Oscilloscope VI, continue to measure the output of the function generator making sure that its output voltage remains at 3.5Vp-p. Use CHANNEL B of the oscilloscope to measure the voltage (magnitude and phase) at point A with respect to ground (Figure 26-3). Record this reading in Table 26-2.

Step #12 Using a Function Generator VI, set the frequency of the V5 to ***2.0 kHz.***
Using an Oscilloscope VI, continue to measure the output of the function generator making sure that its output voltage remains at 3.5Vp-p. Use CHANNEL B of the oscilloscope to measure the voltage (magnitude and phase) at point A with respect to ground (Figure 26-3). Record this reading in Table 26-2.

Step #13 Using a Function Generator VI, set the frequency of the V5 to ***10.0 kHz.***
Using an Oscilloscope VI, continue to measure the output of the function generator making sure that its output voltage remains at 3.5Vp-p. Use CHANNEL B of the oscilloscope to measure the voltage (magnitude and phase) at point A with respect to ground (Figure 26-3). Record this reading in Table 26-2.

This concludes the procedure for this experiment.

Table 26-2

Inductor L1				
Frequency	***Predicted Voltage***		***Measured Voltage***	
	Magnitude [Vp-p]/Phase [degrees]		Magnitude [Vp-p]/Phase [degrees]	
100 Hz				
600 Hz				
2.0k Hz				
10.0k Hz				
Inductor L2				
Frequency	***Predicted Voltage***		***Measured Voltage***	
	Magnitude [Vp-p]/Phase [degrees]		Magnitude [Vp-p]/Phase [degrees]	
100 Hz				
600 Hz				
2.0k Hz				
10.0k Hz				

Questions

1. What is the phase relationship of inductive reactance to resistance?

1. Perform the error calculation for all entries in Table 26-2. Comment on whether the errors are acceptable. Why is the error greater at lower frequencies? (Hint: Consider the effects of inductor winding resistance)

1. What is the phase relationship of voltage across an inductor and current through the inductor?

1. In a series RL circuit, as frequency is increased, does the circuit become more resistive or inductive?

1. Using the results from Table 26-2, demonstrate that Kirchhoff's voltage law applies to series RL circuits.

1. What is the phase relationship between current through a resistor and voltage across a resistor?

Bonus:

1. Use the Impedance Analyzer VI, to measure the total impedance of the circuit shown in Figure 26-4. Record your results in Table 26-3.

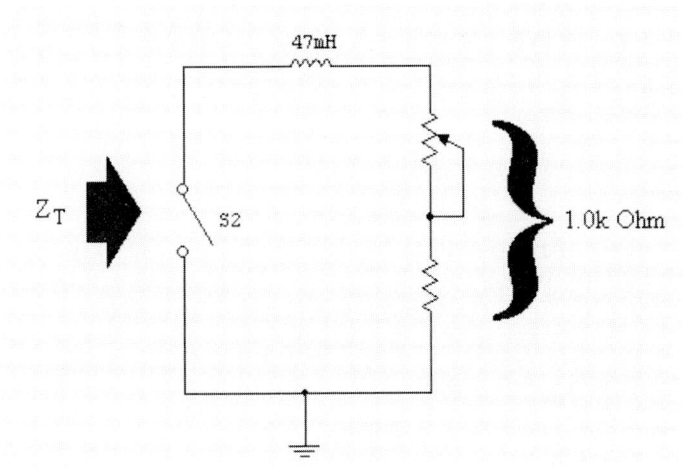

Figure 26-4

Table 26-3

Frequency	***Inductive Reactance***	***Total Impedance***
100 Hz		
600 Hz		
2.0k Hz		
10.0k Hz		

Troubleshooting

Exercise #1

A subwoofer in a given stereo system has an internal resistance of 8 _ and is designed to sound audio signals below 700Hz. The section of a stereo system in Figure 26-5 is not working at its best. There are high frequency signals which are distorted by the subwoofer. Should the inductor be replaced with one having higher or lower inductance value? What is the inductance value in order to establish a maximum power transfer condition at 700Hz.

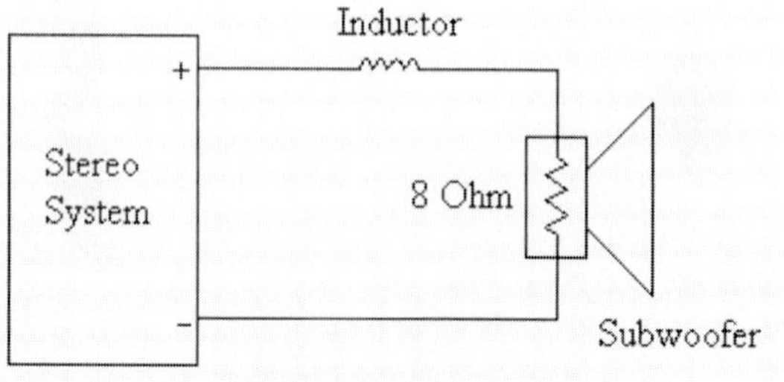

Figure 26-5

27

RC Circuits

Concept

Capacitors resist a change in voltage. The voltage across a capacitor will lag the current through a capacitor by 90°. Capacitors in AC circuits affect the impedance of the circuit by adding capacitive reactance (X_c) in the circuit. Capacitive reactance is calculated by $1/2\pi fC$, where f is frequency in Hertz and C is capacitance in Farads. X_c is inversely related to the frequency of the AC voltage. At high frequencies a capacitor resembles a short. At lower frequencies, capacitors approach an open; that is, capacitors represent an open to DC. X_c is on the Y-axis, of the X-Y coordinate system, causing it to have an angle of -90°. It is an imaginary value.

In series RC circuits, at lower frequencies Xc increases causing the total impedance (Zt) to increase. When X_c equals R the angle of the impedance will be -45°, as shown in Figure 27-1. This will cause the current to have an angle of 45°. When X_c is greater than R, in a series RC circuit, the angle of the impedance will be from -45° to -90°. The current will then be from 45° to 90°.When X_c is less than R in a series, RC circuit, the angle of the impedance will be from 0° to -45°. The current will be from 0° to 45°. Voltage will lag the current by 90°.

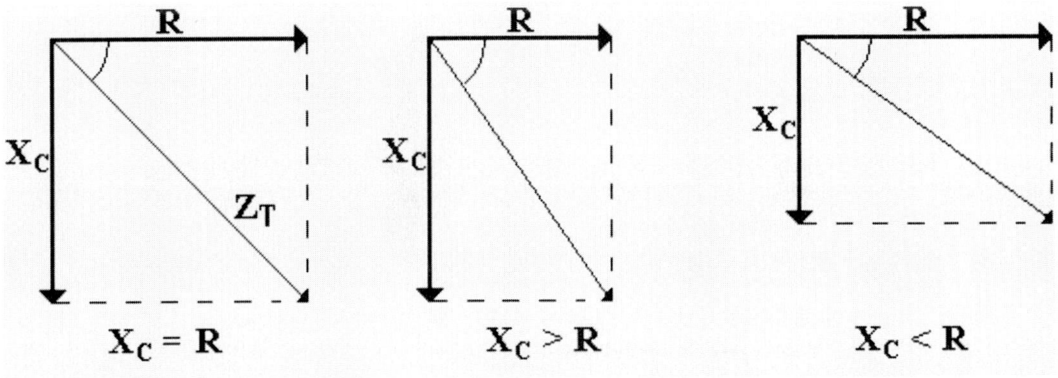

Figure 27-1

Learning Objectives

Upon completion of this lab, the student will be able to:

1. Determine capacitive reactance at various frequencies.

1. Determine total impedance of an RC circuit.
1. Draw the phasor representation of the impedance of an RC circuit.
1. Determine the voltage and current in an RC circuit to include phase angles.
1. Measure the voltage in an RC circuit including phase angles.
1. Draw the phasor representation of the relationship of voltage and current in an RC circuit.

Pre-Lab Procedure

1. Predict the capacitive reactance and total impedance of the circuit shown in Figure 27-2. Record your results in Table 27-1.

1. Draw a phasor diagram of resistance, capacitive reactance and total impedance of the circuit shown in Figure 27-2. Use frequency values from Table 27-1

Table 27-1

Frequency	*Capacitive Reactance*	*Total Impedance*
100 Hz		
600 Hz		
2.0k Hz		
10.0k Hz		

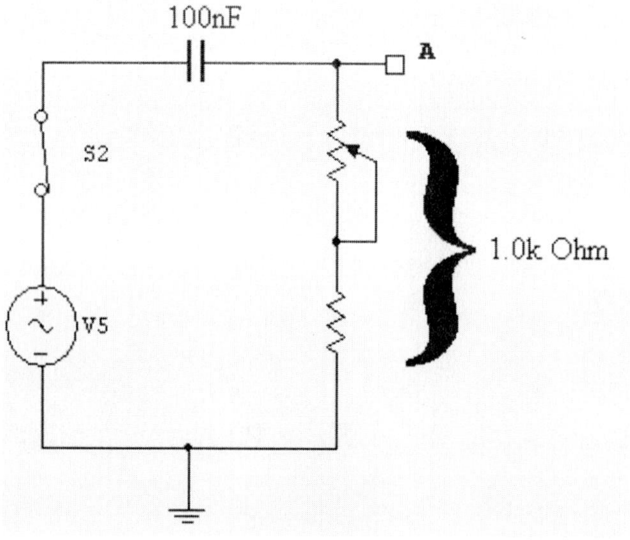

Figure 27-2

1. Predict the voltage (magnitude and phase) at point A of the circuit shown in Figure 27-2. Record your results in Table 27-2. V5=3.5Vp-p.

1. Repeat Pre-Lab procedure #3 for capacitance value of 220nF. Record your results in Table 27-2.

Lab Procedure

Before proceeding further, make sure that your eTCB is working correctly. Using eTCB Diagnostic VI, perform a self-diagnostic of eTCB circuitry.

Turn the switch S1 and S2 OFF in preparation for the lab.

Setup

Step #1 Using the eTCB Configuration VI (AC module), reset all relays.
Using DMM VI, measure the equivalent resistance of POT R34 in series with R35. Adjust its resistance to measure **1.0k Ω.**

Step #2 With all relays still reset, use a DMM VI, to measure the capacitance of C1. Record it below:

Measured Capacitance of C1	

Step #3 With all relays still reset, use a DMM VI, to measure the capacitance of C2. Place jumper **J10** in the B to C position to select capacitor C2. Record its capacitance below:

Measured Capacitance of C2	

Step #4 Using the eTCB Configuration VI, (AC module) configure the circuit in Figure 27-3. Place jumper **J10** back in the A to C position to select capacitor C1.

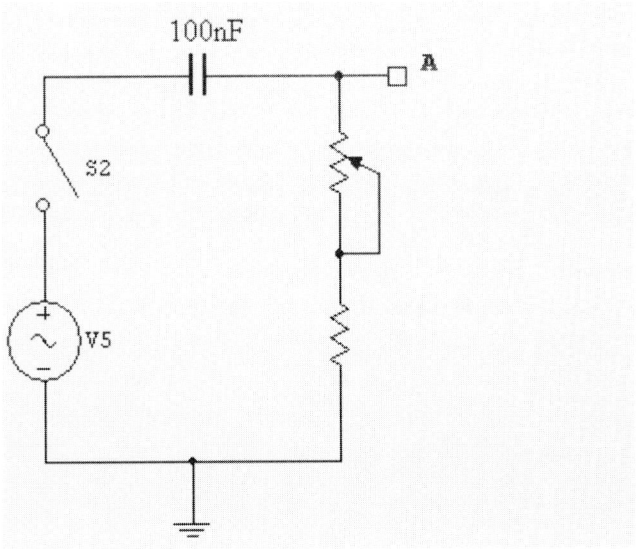

Figure 27-3

Step#4 (cont.) Using a Function Generator VI, set the voltage source V5 to ***1.50V peak*** amplitude at ***100 Hz***. Keep in mind that this setting is just a good indication of what the voltage of V5 is set to. It may not necessarily be the actual voltage of V5. Amplitude of signals from function generators often need to be adjusted (tuned) in order to set their output voltage precisely. Keep the Function Generator VI ***open***; if you close it, its output will reset to 0V.

Step #5 Physically, turn the switch S2 on the eTCB board ON. This will complete your circuit. Use Oscilloscope VI, to measure voltage at TP40 with respect to ground. Display this signal on CHANNEL A. Use the measurement option on the oscilloscope to measure signal parameters on CHANNEL A. Continue to perform this measurement while adjusting the signal amplitude with function generator. Adjust V5 to ***3.5Vp-p.***

Discovery

Step #6 Using an Oscilloscope VI, continue to measure the output of the function generator making sure that its output voltage remains at 3.5Vp-p. You may need to adjust it when it drifts too far away from 3.5Vp-p.

Use CHANNEL B of the oscilloscope to measure the voltage (magnitude and phase) at point A with respect to ground (Figure 27-3). Record this reading in Table 27-2.

Step #7 Using a Function Generator VI, set the frequency of the V5 to ***600 Hz.***
Using an Oscilloscope VI, continue to measure the output of the function generator making sure that its output voltage remains at 3.5Vp-p. Use CHANNEL B of the oscilloscope to measure the voltage (magnitude and phase) at point A with respect to ground (Figure 27-3). Record this reading in Table 27-2.

Step #8 Using a Function Generator VI, set the frequency of the V5 to ***2.0 kHz.***
Using an Oscilloscope VI, continue to measure the output of the function generator making sure that its output voltage remains at 3.5Vp-p. Use CHANNEL B of the oscilloscope to measure the voltage (magnitude and phase) at point A with respect to ground (Figure 27-3). Record this reading in Table 27-2.

Step #9 Using a Function Generator VI, set the frequency of the V5 to ***10.0 kHz.***
Using an Oscilloscope VI, continue to measure the output of the function generator making sure that its output voltage remains at 3.5Vp-p. Use CHANNEL B of the oscilloscope to measure the voltage (magnitude and phase) at point A with respect to ground (Figure 27-3). Record this reading in Table 27-2.

Turn the switch S2 "OFF".

Note: We are now going to examine the effects of different capacitance value on the circuit behavior. We are going to use the same circuit diagram, but rather than choosing to use capacitor C1, you are going to choose ***capacitor C2*** *instead. Placing jumper J10 in the B to C position will select capacitor C2.*

Turn the switch S2 "ON".

Step #10 Using a Function Generator VI, set the frequency of the V5 to ***100 Hz.***

Using an Oscilloscope VI, continue to measure the output of the function generator making sure that its output voltage remains at 3.5Vp-p. Use CHANNEL B of the oscilloscope to measure the voltage (magnitude and phase) at point A with respect to ground (Figure 27-3). Record this reading in Table 27-2.

Step #11 Using a Function Generator VI, set the frequency of the V5 to ***600Hz.***
Using an Oscilloscope VI, continue to measure the output of the function generator making sure that its output voltage remains at 3.5Vp-p. Use CHANNEL B of the oscilloscope to measure the voltage (magnitude and phase) at point A with respect to ground (Figure 27-3). Record this reading in Table 27-2.

Step #12 Using a Function Generator VI, set the frequency of the V5 to ***2.0 kHz.***
Using an Oscilloscope VI, continue to measure the output of the function generator making sure that its output voltage remains at 3.5Vp-p. Use CHANNEL B of the oscilloscope to measure the voltage (magnitude and phase) at point A with respect to ground (Figure 27-3). Record this reading in Table 27-2.

Step #13 Using a Function Generator VI, set the frequency of the V5 to ***10.0 kHz.***
Using an Oscilloscope VI, continue to measure the output of the function generator making sure that its output voltage remains at 3.5Vp-p. Use CHANNEL B of the oscilloscope to measure the voltage (magnitude and phase) at point A with respect to ground (Figure 27-3). Record this reading in Table 27-2.

This concludes the procedure for this experiment.

Table 27-2

Capacitor C1				
Frequency	***Predicted Voltage***		***Measured Voltage***	
	Magnitude [Vp-p]/Phase [degrees]		Magnitude [Vp-p]/Phase [degrees]	
100 Hz				
600 Hz				
2.0k Hz				
10.0k Hz				
Capacitor C2				
Frequency	***Predicted Voltage***		***Measured Voltage***	
	Magnitude [Vp-p]/Phase [degrees]		Magnitude [Vp-p]/Phase [degrees]	
100 Hz				
600 Hz				
2.0k Hz				
10.0k Hz				

Questions

1. What is the phase relationship of capacitive reactance to resistance?

1. Perform the error calculation for all entries in Table 27-2. Comment on whether the errors are acceptable.

1. What is the phase relationship of voltage across a capacitor and current through a capacitor?

1. In a series RC circuit, as frequency is increased does the circuit appear more resistive or capacitive?

1. Using the data from Table 27-2, demonstrate that Kirchhoff's voltage law applies to series RC circuits.

Bonus:

1. Use the Impedance Analyzer VI, to measure the total impedance of the circuit shown in Figure 27-4. Record your results in Table 27-3.

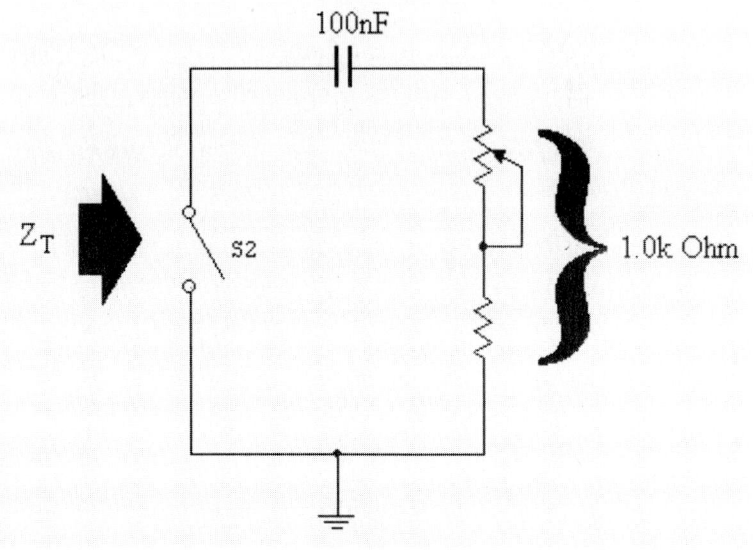

Figure 27-4
S

Table 27-3

Frequency	*Capacitive Reactance*	*Total Impedance*
600 Hz		
2.5k Hz		
6.0k Hz		

28

RLC Series Circuits and Resonance

Problem Statement

Consider a scenario in which you are charged with a task of mapping a strength of a radio signal in a metropolitan area, for a given radio station. This particular radio station is transmitting its signal at 150MHz. Your challenge is to build a circuit that would allow you to detect and gauge the strength of a 150M Hz radio signal. Understanding the behavior of a RLC circuits for input signal with different frequencies will help you in solving your challenge.

Concept

AC circuits with resistors, capacitors, and inductors are known as RLC circuits. The capacitive reactance, inductive reactance, and resistance in the circuit determine the total impedance. Total impedance affects the current in a circuit and the voltage across the components in the circuit. RLC circuits follow the rules of all series and parallel circuits. In series circuits, the largest impedance has the greatest effect on the current in the circuit. In parallel circuits the smallest impedance has the greatest effect on the current in the circuit. In an RLC circuit, there is a frequency where the $X_c = X_L$. This is known as the resonant frequency. At that frequency, X_c and X_L would be equal and 180° out of phase. In a series circuit this would cause the capacitor and inductor combination to appear as a short to the voltage source, causing the resistor to be the only load in the circuit. At this point, impedance would be minimal and current would be maximum. At frequencies above the resonant frequency, X_L would increase and X_c would decrease causing the circuit to be more inductive. At frequencies below the resonant frequencies X_c would increase and X_L would decrease causing the circuit to be more capacitive.

Resonant frequency is determined from the following:

Resonant frequency = 1/(2pisqrt(LC))

Learning Objectives

Upon completion of this lab, the student will be able to:

1. Predict the resonant frequency for an RLC circuit.
1. Predict total impedance of a series RLC circuit.
1. Predict the voltage and current in a series RLC circuit including phase angles.
1. Measure the voltage in a series RLC circuit including phase angles.

Pre-Lab Procedure

1. Predict the resonant frequency of the circuit in Figure 28-2. Record your results in Table 28-1.

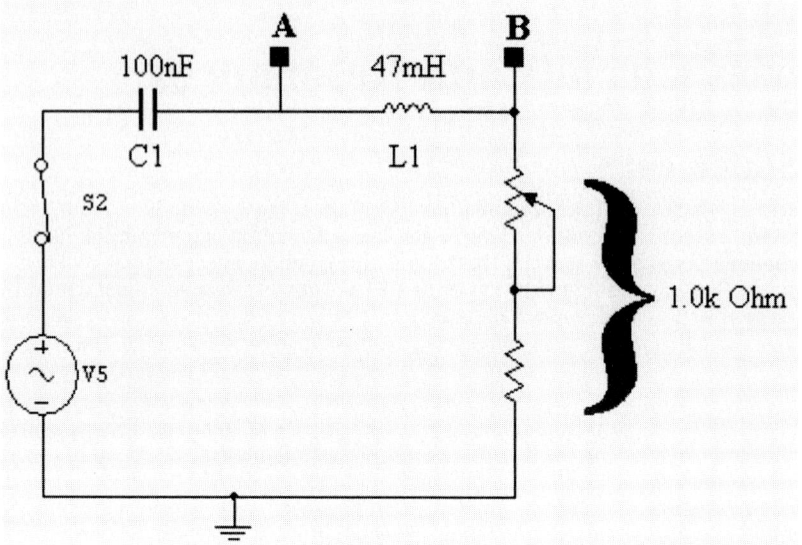

Figure 28-2

1. Predict the capacitive reactance and inductive reactance at the resonant frequency of the circuit in Figure 28-2.

1. Predict the voltages at point A, and B of the circuit in Figure 28-2. Record your results in Table 28-1.

Lab Procedure

Before proceeding further, make sure that your eTCB is working correctly. Using eTCB Diagnostic VI, perform a self-diagnostic of eTCB circuitry.

Turn the switch S1 and S2 OFF in preparation for the lab.

Setup

Step #1 Using eTCB Configuration VI (AC module), reset all relays.
Using DMM VI, measure the equivalent resistance of POT R34 in series with R35.
Adjust its resistance to measure 1.0kΩ.

Step #2 With all relays still reset, use DMM VI to measure capacitance of **C1**. Record it below:

Measured Capacitance of C1	

Step #3 With all relays still reset, use DMM VI, to measure Inductance of **L1**. Record it below:

Measured Inductance of L1	

Step #4 Using eTCB Configuration VI (AC module), configure the circuit in Figure 28-2

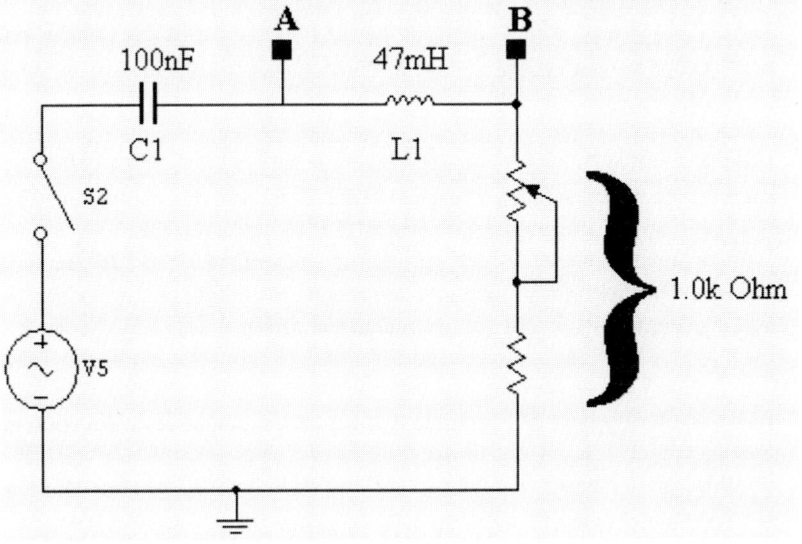

Figure 28-2

Step#4 (cont.) Using a Function Generator VI, set the voltage source V5 to ***1.75V peak*** amplitude at ***500 Hz***. Keep in mind that this setting is just a good indication of what the voltage of V5 is set to. It may not necessarily be the actual voltage of V5. Amplitude of signals from function generators often need to be adjusted (tuned) in order to set their output voltage precisely. Keep the Function Generator VI ***open***; if you close it, its output will reset to 0V.

Step #5 Physically, turn the switch S2 on the eTCB board ON. This will complete your circuit. Use Oscilloscope VI, to measure voltage at ***TP40*** with respect to ground. Display this signal on CHANNEL A. Use the measurement option on the oscilloscope to measure signal parameters on CHANNEL A. Continue to perform this measurement while adjusting the signal amplitude with function generator. Adjust V5 to 3.5Vp-p.

Discovery

Step #6 Using an Oscilloscope VI, continue to measure the output of the function generator making sure that its output voltage remains at 3.5Vp-p. You may need to adjust it when it drifts too far away from 3.5Vp-p. Use CHANNEL B of the oscilloscope to measure the voltage at **Point A** (Figure 28-2). Record this reading in Table 28-1.

Step #7 Using an Oscilloscope VI, continue to measure the output of the function generator making sure that its output voltage remains at 3.5Vp-p. Use CHANNEL B of the oscilloscope to measure the voltage at **Point B** (Figure 28-2). Record this reading in Table 28-1.

Step #8 Using a Function Generator VI, set the frequency of the V5 to ***1.0k Hz.***
Using an Oscilloscope VI, continue to measure the output of the function generator making sure that its output voltage remains at 3.5Vp-p. Use CHANNEL B of the oscilloscope to measure the voltage at **Point A.** Record this reading in Table 28-1.

Step #9 Using an Oscilloscope VI, continue to measure the output of the function generator making sure that its output voltage remains at 3.5Vp-p. Use CHANNEL B of the oscilloscope to measure the voltage at **Point B.** Record this reading in Table 28-1.

Step #10 Using a Function Generator VI, set the frequency of the V5 to ***2.0k Hz.***
Using an Oscilloscope VI, continue to measure the output of the function generator making sure that its output voltage remains at 3.5Vp-p. Use CHANNEL B of the oscilloscope to measure the voltage at **Point A.** Record this reading in Table 28-1.

Step #11 Using an Oscilloscope VI, continue to measure the output of the function generator making sure that its output voltage remains at 3.5Vp-p. Use CHANNEL B of the oscilloscope to measure the voltage at **Point B.** Record this reading in Table 28-1.

Step #12 Using a Function Generator VI, set the frequency of the V5 to ***the predicted resonant frequency.***

Using an Oscilloscope VI, continue to measure the output of the function generator making sure that its output voltage remains at 3.5Vp-p. Use CHANNEL B of the oscilloscope to measure the voltage at **Point A.** Record this reading in Table 28-1.

Step #13 Using an Oscilloscope VI, continue to measure the output of the function generator making sure that its output voltage remains at 3.5Vp-p. Use CHANNEL B of the oscilloscope to measure the voltage at **Point B.** Record this reading in Table 28-1.

Step #14 Using a Function Generator VI, set the frequency of the V5 to ***3.0k Hz.***
Using an Oscilloscope VI, continue to measure the output of the function generator making sure that its output voltage remains at 3.5Vp-p. Use CHANNEL B of the oscilloscope to measure the voltage at **Point A** Record this reading in Table 28-1.

Step #15 Using an Oscilloscope VI, continue to measure the output of the function generator making sure that its output voltage remains at 3.5Vp-p. Use CHANNEL B of the oscilloscope to measure the voltage at **Point A.** Record this reading in Table 28-1.

Step #16 Using a Function Generator VI, set the frequency of the V5 to ***6.0k Hz.***

Using an Oscilloscope VI, continue to measure the output of the function generator making sure that its output voltage remains at 3.5Vp-p. Use CHANNEL B of the oscilloscope to measure the voltage at **Point A.** Record this reading in Table 28-1.

Step #17 Using an Oscilloscope VI, continue to measure the output of the function generator making sure that its output voltage remains at 3.5Vp-p. Use CHANNEL B of the oscilloscope to measure the voltage at **Point B.** Record this reading in Table 28-1.

Step #18 Using a Function Generator VI, set the frequency of the V5 to ***10.0k Hz***.

Using an Oscilloscope VI, continue to measure the output of the function generator making sure that its output voltage remains at 3.5Vp-p. Use CHANNEL B of the oscilloscope to measure the voltage at **Point A.** Record this reading in Table 28-1.

Step #19 Using an Oscilloscope VI, continue to measure the output of the function generator making sure that its output voltage remains at 3.5Vp-p. Use CHANNEL B of the oscilloscope to measure the voltage at **Point B.** Record this reading in Table 28-1.

This concludes the procedure for this experiment.

Table 28-1

Predicted Resonance Frequency	***Estimated Resonance Frequency***

Frequency	*Predicted **Point A** Voltage [Vp-p/Angle]*		*Measured **Point A** Voltage [Vp-p/Angle]*	
500 Hz				
1.0k Hz				
2.0k Hz				
Resonant Frequency				
3.0k Hz				
6.0k Hz				
10.0k Hz				

Frequency	*Predicted **Point B** Voltage [Vp-p/Angle]*		*Measured **Point B** Voltage [Vp-p/Angle]*	
500 Hz				
1.0k Hz				
2.0k Hz				
Resonant Frequency				
3.0k Hz				
6.0k Hz				
10.0k Hz				

Questions

1. What occurs at the resonant frequency?

1. Perform the error calculation for all entries in Table 28-1. Comment on whether the errors are acceptable.

1. Below the resonant frequency, is the series RLC circuit more capacitive or inductive?

1. At frequencies above resonant frequency, is the series RLC circuit more capacitive or inductive?

1. Using the data from Table 28-1, demonstrate that Kirchhoff's voltage law applies to series RLC circuits.

1. When is current flow the greatest in a series RLC circuit?

1. How can the RLC circuit be used to solve this experiment's challenge posed in the Problem Statement?

Bonus:

Use the Impedance Analyzer VI, to measure the total impedance of the circuit shown in Figure 28-4. Record your results in Table 28-2.

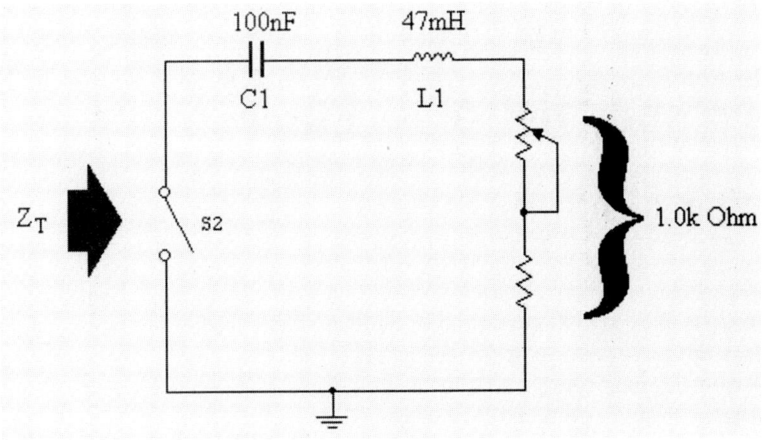

Figure 28-4

Table 28-2

Frequency	***Reactance***	***Total Impedance***
500 Hz		
1.0k Hz		
2.0k Hz		
Resonant Frequency		
3.0k Hz		
6.0k Hz		
10.0k Hz		

29

Low-Pass and High-Pass Filters

Problem Statement

There often exists a situation where electrical signals with a specific range of frequencies are of special interest. In some cases those signals are desirable, in other situations they are unwanted. High-pass filters allow frequencies **above** a certain cutoff frequency (fc) to pass and reject all others. Low-pass filters allow all frequencies **below** a certain cutoff frequency and reject all frequencies above the fc. Let's examine a situation with a stereo system in which we want take advantage of a particularly good low frequency (base) response of a subwoofer, and a particularly good high frequency response of a tweeter speaker. In this scenario we would like to take advantage of a low-pass filter function, and pass the low frequencies (base) to a subwoofer. We would also like to take advantage of a high-pass filter function and pass the high frequencies to tweeter speakers.

Concept

Filters are circuits that are designed to allow signals of certain frequencies to pass to a load, and to reject all other frequencies. Because both capacitors and inductors respond to frequencies, they are used to design and construct filters. A low-pass filter may be constructed from a series RC or RL circuit:

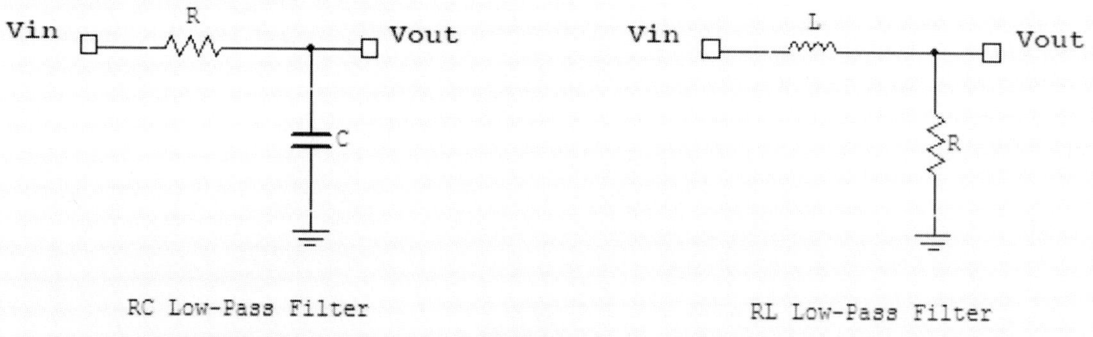

Figure 29-1

Looking at the RC filter in Figure 29-1, it can be determined the output voltage is developed across the capacitor. Xc increases as frequency drops. As the capacitive reactance Xc approaches an open, more of the voltage is developed across the capacitor and less across the resistor. The cutoff frequency is achieved when Xc = R. At this point the output voltage will be 70.7 percent of the input voltage.

Considering the RL low-pass filter in Figure 29-1, the output voltage is developed across a resistor, which is in series with an inductor. As frequency increases XL increases. This causes an increase in voltage across the inductor and decrease in voltage across the resistor. At the cutoff frequency XL = R, the voltage across the resistor will equal 70.7 percent of the input voltage.

At frequencies above cutoff, the signals are rejected. In the RC low-pass filter, the voltage is shorted to ground. In the RL circuit, the voltage is developed across the inductor, which approaches an open as the frequency is increased.

The cutoff frequency for the RC circuit is determined by the following equation:

$$Fc=1/(2\pi RC)$$

The cutoff frequency for the RL circuit is determined by the following equation:

$$Fc=R/(2\pi L)$$

A high-pass filter may also be constructed from a series RC or RL circuit as seen in Figure 29-2:

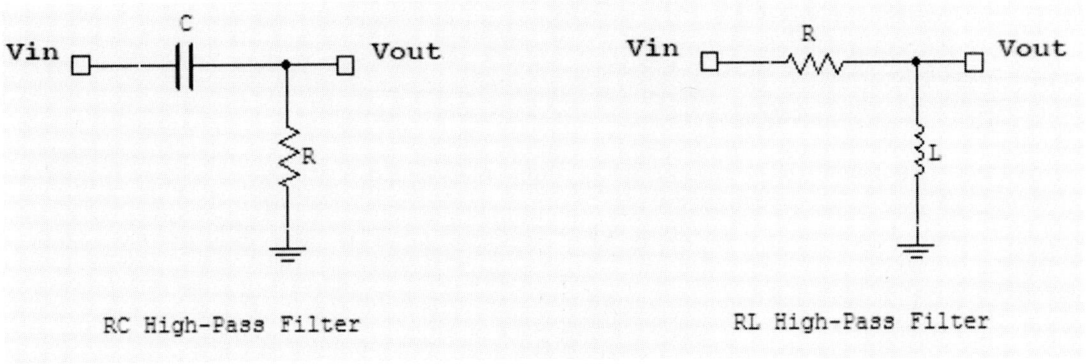

Figure 29-2

The RC circuit in Figure 29-2 causes the output voltage to be developed across the resistor in series with the capacitor. As frequencies increase, the capacitor becomes a short, causing most of the signal to be developed across the resistor. The cutoff frequency is reached when $X_C = R$. At this point 70.7 percent of the applied voltage is developed across the resistor. All frequencies below fc are rejected, all frequencies above fc are passed.

The RL circuit in Figure 29-2 causes the output voltage to be developed across the inductor. As the frequency increases, X_L increases. At fc $X_L = R$, 70.7 percent of the applied voltage will then be developed across the inductor. The filter passes all frequencies above fc and rejects all frequencies below fc. The equation for the cutoff frequency is the same for high-pass and low-pass filters.

Learning Objectives

Upon completion of this lab, the student will be able to:

1. Identify and configure a low-pass and high-pass filter.
1. Graph the frequency response of a filter.
1. Determine the cut-off frequency of a filter.

Pre-Lab Procedure

1. Predict the cut-off frequency of the filter in Figure 29-3. Record your results in Table 29-1.

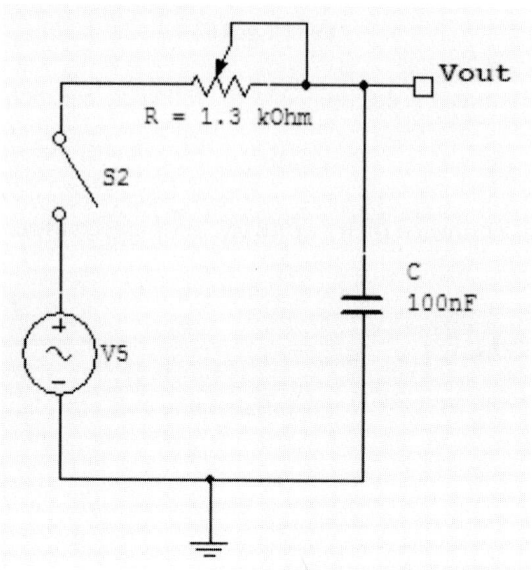

Figure 29-3

1. Predict the cut-off frequency of the filter in Figure 29-4. Record your results in Table 29-3.

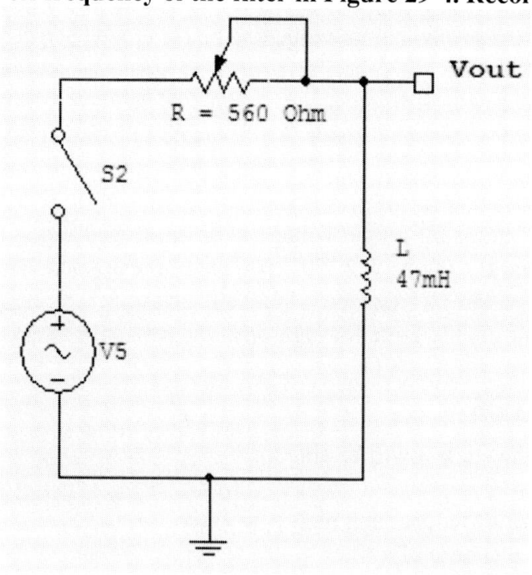

Figure 29-4

Lab Procedure

Before proceeding further, make sure that your eTCB is working correctly. Using eTCB Diagnostic VI, perform a self-diagnostic of eTCB circuitry.

Turn the switch S1 and S2 OFF in preparation for the lab.

Setup

Step #1 — Using the eTCB Configuration VI, (AC module), reset all relays.
Using DMM VI, measure resistance of POT R30. Adjust its resistance to measure 1.3KΩ.

Step #2 — With all relays still reset, use DMM VI, to measure capacitance of **C3**. Record it below:

Measured Capacitance of C3	

Step #3 — With all relays still reset, use DMM VI, to measure inductance of **L3**. Record it below:

Measured Inductance of L3	

Step #4 — Using the eTCB Configuration VI, (AC module) configure the circuit in Figure 29-5

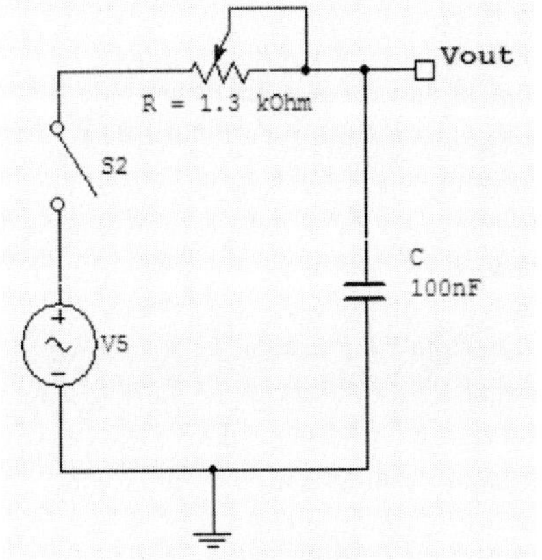

Figure 29-5

Step#4 (cont.) — Using a Function Generator VI, set the voltage source V5 to ***1.75V peak*** amplitude at ***200 Hz***. Keep in mind that this setting is just a good indication of what the voltage of V5 is set to. It may not necessarily be the actual voltage of V5. Amplitude of signals from function generators, often need to be adjusted (tuned) in order to set their output voltage precisely. Keep the Function Generator VI ***open***; if you close it, its output will reset to 0V.

Step #5 — Physically, turn the switch S2 on the eTCB board ON. This will complete your circuit. Use Oscilloscope VI, to measure voltage at TP40 with respect to ground. Display this signal on CHANNEL A. Use the measurement option on the oscilloscope to measure signal parameters on CHANNEL A. Continue to perform this measurement while adjusting the signal amplitude with function generator. Adjust V5 to 3.5Vp-p.

Discovery

Low-Pass Filter

Step #6 Using an Oscilloscope VI, continue to measure the output of the function generator making sure that its output voltage remains at 3.5Vp-p. You may need to adjust it when it drifts too far away from 3.5Vp-p. Use CHANNEL B of the oscilloscope to measure output voltage Vout (Figure 29-5). Record this reading in Table 29-2.

Step #7 Using a Function Generator VI, set the frequency of the V5 to ***600 Hz.***
Using an Oscilloscope VI, continue to measure the output of the function generator making sure that its output voltage remains at 3.5Vp-p. Use CHANNEL B of the oscilloscope to measure output voltage Vout (Figure 29-5). Record this reading in Table 29-2.

Step #8 Using a Function Generator VI, set the frequency of the V5 to ***1.0k Hz.***
Using an Oscilloscope VI, continue to measure the output of the function generator making sure that its output voltage remains at 3.5Vp-p. Use CHANNEL B of the oscilloscope to measure output voltage Vout (Figure 29-5). Record this reading in Table 29-2.

Step #9 Using a Function Generator VI, set the frequency of the V5 to ***1.3k Hz.***
Using an Oscilloscope VI, continue to measure the output of the function generator making sure that its output voltage remains at 3.5Vp-p. Use CHANNEL B of the oscilloscope to measure output voltage Vout (Figure 29-5). Record this reading in Table 29-2.

Step #10 Using a Function Generator VI, set the frequency of the V5 to ***2.5k Hz.***
Using an Oscilloscope VI, continue to measure the output of the function generator making sure that its output voltage remains at 3.5Vp-p. Use CHANNEL B of the oscilloscope to measure output voltage Vout (Figure 29-5). Record this reading in Table 29-2.

Step #11 Using a Function Generator VI, set the frequency of the V5 to ***5.0k Hz.***
Using an Oscilloscope VI, continue to measure the output of the function generator making sure that its output voltage remains at 3.5Vp-p. Use CHANNEL B of the oscilloscope to measure output voltage Vout (Figure 29-5). Record this reading in Table 29-2.

High-Pass Filter

Step #12 Using the eTCB Configuration VI, reset all relays.
Using DMM VI, measure resistance of POT R30. Adjust its resistance to measure 560Ω.

Step #13 Using the eTCB Configuration VI, configure the circuit in Figure 29-6.

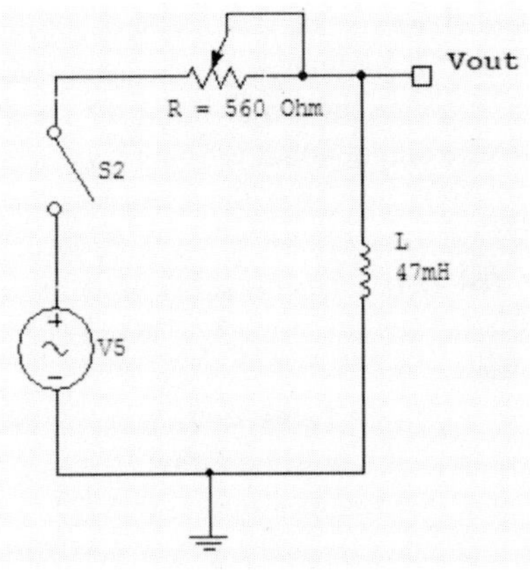

Figure 29-6

Using a Function Generator VI, set the voltage source V5 to ***1.75V peak*** amplitude at ***200 Hz***.

Using an Oscilloscope VI, continue to measure the output of the function generator making sure that its output voltage remains at 3.5Vp-p. Use CHANNEL B of the oscilloscope to measure output voltage Vout (Figure 29-6). Record this reading in Table 29-4.

Step #14 Using a Function Generator VI, set the frequency of the V5 to ***1.0k Hz***. Using an Oscilloscope VI, continue to measure the output of the function generator making sure that its output voltage remains at 3.5Vp-p. Use CHANNEL B of the oscilloscope to measure output voltage Vout (Figure 29-6). Record this reading in Table 29-4.

Step #15 Using a Function Generator VI, set the frequency of the V5 to ***2.0k Hz***. Using an Oscilloscope VI, continue to measure the output of the function generator making sure that its output voltage remains at 3.5Vp-p. Use CHANNEL B of the oscilloscope to measure output voltage Vout (Figure 29-6). Record this reading in Table 29-4.

Step #16 Using a Function Generator VI, set the frequency of the V5 to ***2.5k Hz***. Using an Oscilloscope VI, continue to measure the output of the function generator making sure that its output voltage remains at 3.5Vp-p. Use CHANNEL B of the oscilloscope to measure output voltage Vout (Figure 29-6). Record this reading in Table 29-4.

Step #17 Using a Function Generator VI, set the frequency of the V5 to ***3.0k Hz***. Using an Oscilloscope VI, continue to measure the output of the function generator making sure that its output voltage remains at 3.5Vp-p. Use CHANNEL B of the oscilloscope to measure output voltage Vout (Figure 29-6). Record this reading in Table 29-4.

Step #18 Using a Function Generator VI, set the frequency of the V5 to ***6.0k Hz***.
Using an Oscilloscope VI, continue to measure the output of the function generator making sure that its output voltage remains at 3.5Vp-p. Use CHANNEL B of the oscilloscope to measure output voltage Vout (Figure 29-6). Record this reading in Table 29-4.

This concludes the procedure for this experiment.

Table 29-1
Low-Pass Filter

Predicted Cut-off Frequency	*Estimated Cut-off Frequency*

Table 29-2

Frequency	*Measured Voltage [Vp-p]*
200 Hz	
600 Hz	
1.0k Hz	
1.3k Hz	
2.5k Hz	
5.0k Hz	

Table 29-3
High-Pass Filter

Predicted Cut-off Frequency	*Estimated Cut-off Frequency*

Table 29-4

Frequency	*Measured Voltage [Vp-p]*
200 Hz	
1.0k Hz	
2.0k Hz	
2.5k Hz	
3.0k Hz	
6.0k Hz	

Questions

1. What is the equation for cut-off frequency for a high- and a low-pass filter?

1. What does the reactance equal in a high- and a low-pass RC and RL filter at cut-off frequency?

1. Use the data from Table 29-2 to graph the frequency response of low-pass filter. Use semi-log graph paper to complete this question.

1. Considering values recorded in Table 29-2, estimate the value of the cut-off frequency. Record your estimations in Table 29-1. Comment on the accuracy of the estimated values.

1. Use the data from Table 29-4 to graph the frequency response of high-pass filter. Use semi-log graph paper to complete this question.

1. Considering values recorded in Table 29-4, estimate the value of the cut-off frequency. Record your estimations in Table 29-3. Comment on the accuracy of the estimated values.

Bonus:

1. Suggest a way of using both a low-pass and a high-pass filter to build a band-pass filter. A band-pass filter will pass frequency band between with 2.5 kHz and 4.0 kHz. Draw a schematic for the desired filter. Specify the cut-off frequency for both low-pass and high-pass filter.

30

Band-Pass and Band-Stop Filters

Problem Statement

There often exists a situation where electrical signals with a specific range of frequencies are of special interest. In some cases those signals are desirable; in others they are unwanted. Circuits that can selectively pass or reject signals with a specific range of frequencies are called band-pass and band-stop, respectively. Let's examine a situation in which we want to listen to a particular radio station. In this scenario, we would like to take advantage of band-pass filter function and pass the signals of a desirable radio station and reject the rest. Let's also examine a situation where a switching power supply is producing unwanted signals relating to its switching frequency. In this scenario, we would like to take advantage of band-stop filter function and reject the undesirable signals.

Concept

Besides high-pass and low-pass filters, there are two other primary types of filters. They are band-pass and band-stop. The band-stop is also known as the notch filter. Band-pass and notch filters are designed and constructed from RLC circuits. When the output signal is developed across the resistor R, the filter is a band-pass. When the output is across the LC portion of the circuit the filter is a notch filter. These circuits are shown in Figure 30-1.

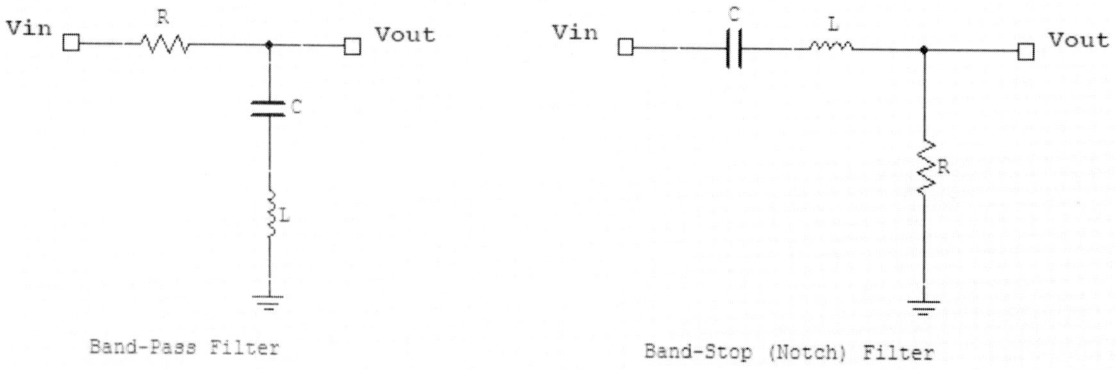

Figure 30-1

The band-pass filter allows frequencies within a defined bandwidth to be passed. All frequencies above and below the bandwidth will be rejected. At the resonant frequency the LC portion of the circuit appear as a short. This causes the entire signal to be developed across the resistor. As the input frequency drops below the resonant frequency, X_c increases and X_L decreases. At some frequency below the resonance, X_c will equal R. This will be the lower cut-off frequency. At frequencies above resonance, X_L increases and X_c decreases. At some frequency above resonance X_L will equal R. This frequency is the upper cut-off frequency. The range of frequencies between the upper and lower cut-offs is the bandwidth.

The band-stop filter, or notch filter, rejects all frequencies within the bandwidth. All frequencies below the lower cutoff and above the upper cutoff are passed. The output of the notch filter is across the LC portion of the circuit. At resonance the LC portion of the circuit represents a short, causing the output to be approximately zero volts. At frequencies below resonance, X_c increases and X_L decreases. At the lower cut-off frequency, X_c equals R. The circuit resembles a low-pass filter. At frequencies above resonance, X_L increases and X_c decreases. At the upper cutoff frequency, X_L equals R. The circuit resembles a high-pass filter. The bandwidth of rejected frequencies is the difference between the upper cut-off and lower cut-off frequencies. There is a characteristic of these filters known as the quality factor, or Q, of the circuit. The Q of the circuit is determined by dividing X_L at resonance by the resistance of the circuit. The Q of the circuit is used to determine the bandwidth of the circuit. The resonant frequency divided by the Q of the circuit provides the bandwidth. The bandwidth is centered on the resonant frequency. The upper and lower cut-off frequencies are determined by adding and subtracting half of the bandwidth to and from the resonant frequency.

Learning Objectives

Upon completion of this lab, the student will be able to:

1. Identify and configure a band-pass and band-stop filter.
1. Graph the frequency response of a filter.
1. Determine the bandwidth of a filter.

Pre-Lab Procedure

1. Predict the resonance frequency, bandwidth, upper, and lower cut-off frequencies of the filter in Figure 30-2. Record your results in Table 30-1.

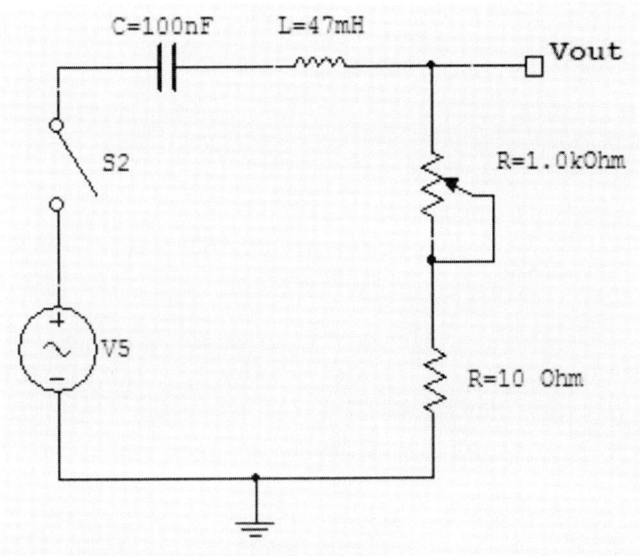

Figure 30-2

1. Predict the resonant frequency, bandwidth, upper, and lower cut-off frequencies of the filter in Figure 30-3. Record your results in Table 30-3.

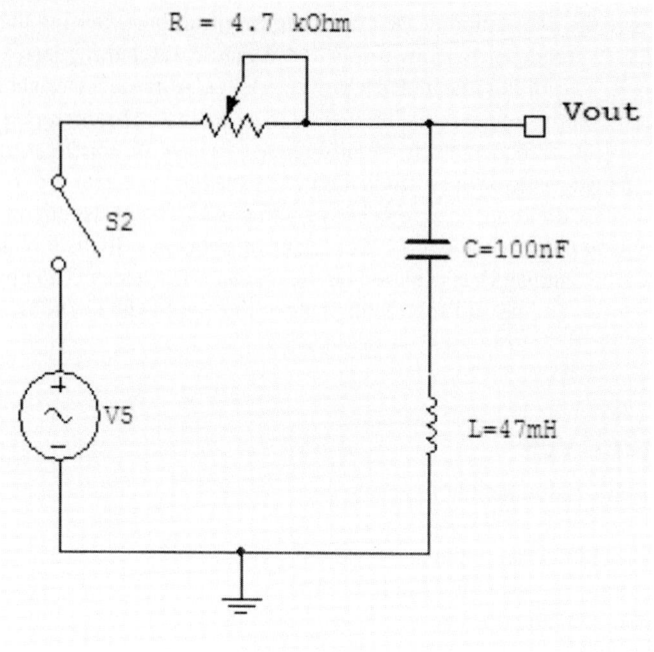

Figure 30-3

Lab Procedure

Before proceeding further, make sure that your eTCB is working correctly. Using eTCB Diagnostic VI, perform a self-diagnostic of eTCB circuitry.

Turn the switch S1 and S2 OFF in preparation for the lab.

Setup

Step #1 Using the eTCB Configuration VI, (AC module), reset all relays. Using a DMM VI, measure resistance of POT R34 in series with R35. Adjust their equivalent resistance to measure 1.0kΩ.

Step #2 With all relays still reset, use DMM VI, to measure resistance of POT R30. Adjust it to measure 4.7kΩ.

Step #3 Using the eTCB Configuration VI, (AC Module) configure the circuit in Figure 30-4

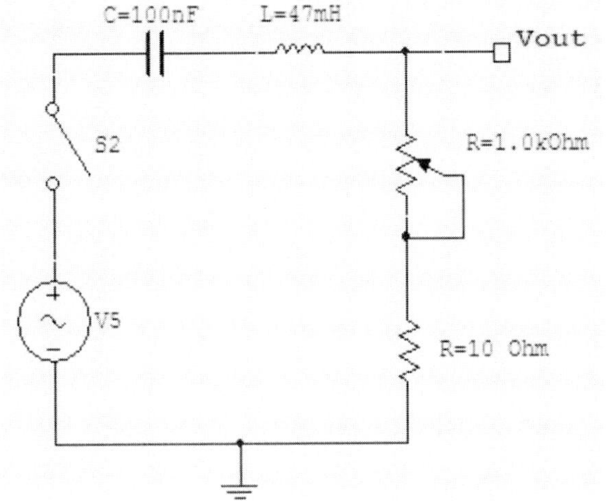

Figure 30-4

Step#3 (cont.) Using a Function Generator VI, set voltage source V5 to ***1.75V*** peak amplitude at ***300 Hz***. Keep in mind that this setting is just a good indication of what the voltage of V5 is set to. It may not necessarily be the actual voltage of V5. Amplitude of signals from function generators often need to be adjusted (tuned) in order to set their output voltage precisely. Keep the Function Generator VI ***open***; if you close it, its output will reset to 0V.

Step #4 Physically, turn the switch S2 on the eTCB board ON. This will complete your circuit. Use Oscilloscope VI, to measure voltage at TP40 with respect to ground. Display this signal on CHANNEL A. Use the measurement option on the oscilloscope to measure signal parameters on CHANNEL A. Continue to perform this measurement while adjusting the signal amplitude with function generator. Adjust V5 to 3.5Vp-p.

Discovery

Band-Pass Filter

Step #5 Using an Oscilloscope VI, continue to measure the output of the function generator making sure that its output voltage remains at 3.5Vp-p. You may need to adjust it when it drifts too far away from 3.5Vp-p. Use CHANNEL B of the oscilloscope to measure output voltage Vout (Figure 30-4). Record this reading in Table 30-2.

Step #6 Using a Function Generator VI, set the frequency of the V5 to ***1.0k Hz***. Using an Oscilloscope VI, continue to measure the output of the function generator making sure that its output voltage remains at 3.5Vp-p. Use CHANNEL B of the oscilloscope to measure output voltage Vout (Figure 30-4). Record this reading in Table 30-2.

Step #7 Using a Function Generator VI, set the frequency of the V5 to ***1.5k Hz***.
Using an Oscilloscope VI, continue to measure the output of the function generator making sure that its output voltage remains at 3.5Vp-p. Use CHANNEL B of the oscilloscope to measure output voltage Vout (Figure 30-4). Record this reading in Table 30-2.

Step #8 Using a Function Generator VI, set the frequency of the V5 to ***2.0k Hz***.
Using an Oscilloscope VI, continue to measure the output of the function generator making sure that its output voltage remains at 3.5Vp-p. Use CHANNEL B of the oscilloscope to measure output voltage Vout (Figure 30-4). Record this reading in Table 30-2.

Step #9 Using a Function Generator VI, set the frequency of the V5 to ***2.5k Hz***.
Using an Oscilloscope VI, continue to measure the output of the function generator making sure that its output voltage remains at 3.5Vp-p. Use CHANNEL B of the oscilloscope to measure output voltage Vout (Figure 30-4). Record this reading in Table 30-2.

Step #10 Using a Function Generator VI, set the frequency of the V5 to ***5.0k Hz***.
Using an Oscilloscope VI, continue to measure the output of the function generator making sure that its output voltage remains at 3.5Vp-p. Use CHANNEL B of the oscilloscope to measure output voltage Vout (Figure 30-4). Record this reading in Table 30-2.

Step #11 Using a Function Generator VI, set the frequency of the V5 to ***10k Hz***.
Using an Oscilloscope VI, continue to measure the output of the function generator making sure that its output voltage remains at 3.5Vp-p. Use CHANNEL B of the oscilloscope to measure output voltage Vout (Figure 30-4). Record this reading in Table 30-2.

Band-Stop Filter

Step #12 Using the eTCB Configuration VI,, configure the circuit in Figure 30-5

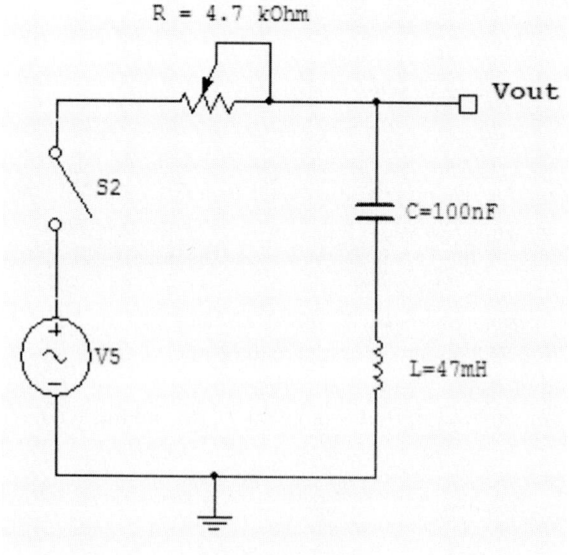

Figure 30-5

Step #13 Using a Function Generator VI, set voltage source V5 to ***1.75V*** peak amplitude at ***100 Hz*** Using an Oscilloscope VI, continue to measure the output of the function generator making sure that its output voltage remains at 3.5Vp-p. Use CHANNEL B of the oscilloscope to measure the output voltage Vout (Figure 30-3). Record this reading in Table 30-4.

Step #14 Using a Function Generator VI, set the frequency of the V5 to ***600 Hz***. Using an Oscilloscope VI, continue to measure the output of the function generator making sure that its output voltage remains at 3.5Vp-p. Use CHANNEL B of the oscilloscope to measure the output voltage Vout (Figure 30-5). Record this reading in Table 30-4.

Step #15 Using a Function Generator VI, set the frequency of the V5 to ***1.0k Hz.*** Using an Oscilloscope VI, continue to measure the output of the function generator making sure that its output voltage remains at 3.5Vp-p. Use CHANNEL B of the oscilloscope to measure the output voltage Vout (Figure 30-5). Record this reading in Table 30-4.

Step #16 Using a Function Generator VI, set the frequency of the V5 to ***2.0k Hz***. Using an Oscilloscope VI, continue to measure the output of the function generator making sure that its output voltage remains at 3.5Vp-p. Use CHANNEL B of the oscilloscope to measure the output voltage Vout (Figure 30-5). Record this reading in Table 30-4.

Step #17 Using a Function Generator VI, set the frequency of the V5 to ***9.0k Hz***. Using an Oscilloscope VI, continue to measure the output of the function generator making sure that its output voltage remains at 3.5Vp-p. Use CHANNEL B of the oscilloscope to measure the output voltage Vout (Figure 30-5). Record this reading in Table 30-4.

Step #18 Using a Function Generator VI, set the frequency of the V5 to ***20k Hz***. Using an Oscilloscope VI, continue to measure the output of the function generator making sure that its output voltage remains at 3.5Vp-p. Use CHANNEL B of the oscilloscope to measure the output voltage Vout (Figure 30-5). Record this reading in Table 30-4.

This concludes the procedure for this experiment.

Table 30-1

Band-Pass Filter

Predicted Resonant Frequency	Estimated Resonant Frequency
Predicted Bandwidth	**Estimated Bandwidth**
Predicted Lower Cut-off Frequency	**Estimated Lower Cut-off Frequency**
Predicted Upper Cut-off Frequency	**Estimated Upper Cut-off Frequency**

Table 30-2

Frequency	*Measured Voltage [Vp-p]*
300 Hz	
1.0k Hz	
1.5k Hz	
2.0k Hz	
2.5k Hz	
5.0k Hz	
10k Hz	

Table 30-3

Band-Stop Filter

Predicted Resonant Frequency	Estimated Resonant Frequency
Predicted Bandwidth	**Estimated Bandwidth**
Predicted Lower Cut-off Frequency	**Estimated Lower Cut-off Frequency**
Predicted Upper Cut-off Frequency	**Estimated Upper Cut-off Frequency**

Table 30-4

Frequency	*Measured Voltage [Vp-p]*
100 Hz	
600 Hz	
1.0k Hz	
2.0k Hz	
9.0k Hz	
20.0k Hz	

Questions

1. What is the relationship of reactance to resistance in band-pass and band-stop filter at upper cut-off frequency?

1. What is the relationship of reactance to resistance in band-pass and band-stop filter at lower cut-off frequency?

1. Use the data from Table 30-2 to graph the frequency response of band-pass filter. Use semi-log graph paper to complete this question.

1. Considering values recorded in Table 30-2, estimate the value of resonant frequency, bandwidth, upper and lower cutoff frequencies of this band-pass filter. Record your estimations in Table 30-1. Comment on the accuracy of the estimated values.

1. Use the data from Table 30-4 to graph the frequency response of band-pass filter. Use semi-log graph paper to complete this question.

1. Considering values recorded in Table 30-4, estimate the value of resonant frequency, bandwidth, upper and lower cutoff frequencies of this band-pass filter. Record your estimations in Table 30-3. Comment on the accuracy of the estimated values.

Bonus:

1. When the value of resistor in the band-pass filter is changed, how does it affect the filter's frequency response?

31

Transformer

Problem Statement

Consider a scenario in which an audio amplifier's output impedance does not match the impedance of a speaker. We know that if the impedances are mismatched the power transfer from the audio amplifier to the speaker will not be at its maximum. Your challenge is to build a circuit with an audio transformer to achieve a maximum power transfer from the audio amplifier to its speaker. Understanding the characteristics of the transformer circuits will help you in solving this challenge.

Concept

Voltage transformers are designed to raise or lower AC voltage. Transformers have an input side and an output side. The input side is called the primary side of the transformer; the output is the secondary. When the voltage on the primary, or input side, is larger than the voltage on the output side, the secondary side, the transformer is a step-down transformer. When the voltage on the primary side is less than the secondary voltage the transformer is a step-up type. Transformers that have primary voltage equal to their secondary voltage are isolation transformers. Isolation transformers are used to isolate common system grounds from each other. Transformers operate on the principle of mutual inductance. As current flows through the primary side winding of the transformer, a magnetic field expands from the winding. The expanding magnetic field cuts across the secondary winding of the transformer, creating a difference of potential across the winding. The more windings in the secondary coil, the greater the difference of potential across the secondary winding. Transformers are described with respect to the number of turns of the primary winding divided by the number of turns of the secondary winding. The same ratio may be determined by dividing the primary voltage by the secondary voltage. For example, a transformer with a turns ratio of 10/1 or 10:1, will produce one volt on the secondary for every 10 volts on the primary. The reverse is also true. Transformers have a power rating, referred to as the VA rating. This is an RMS value. Ideally, the power on the secondary side of a transformer will equal the power on the primary side. This means that a step-down transformer will cause the secondary voltage to be reduced by the turns ratio, but the secondary current will be increased by the reciprocal of the turns ratio. The reverse is also true for the step-up transformer.

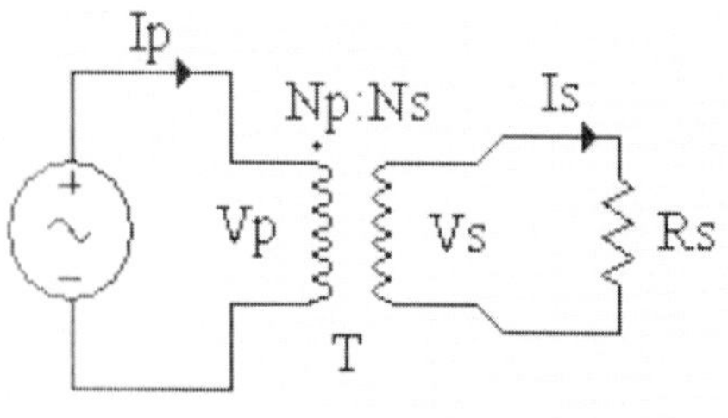

Turns ratio $n=Np/Ns$ $Np/Ns=Vp/Vs$ $Np/Ns=Is/Ip$ $(Np/Ns)2=Rp/Rs$

Learning Objectives

Upon completion of this lab, the student will be able to:

1. Determine the primary and secondary voltage.
1. Determine the primary and secondary power.
1. Determine the equivalent primary resistance.

Measure the primary and secondary voltage.

Pre-Lab Procedure

1. Given the AC voltage V5=1.0Vrms, predict the power dissipation in resistor R_L1 in circuit shown in Figure 31-1. Record your result in Table 31-1.

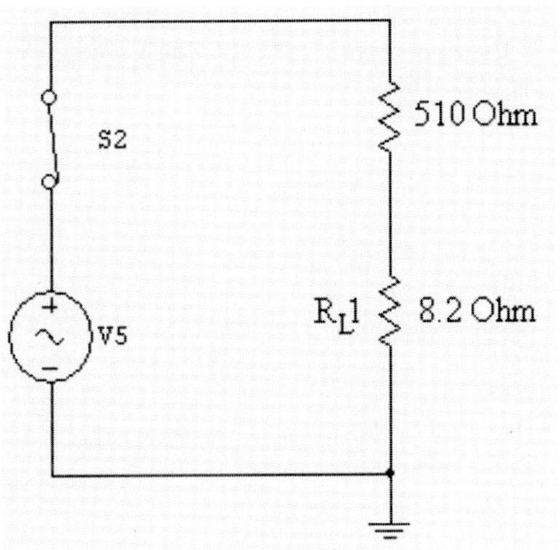

Figure 31-1

1. Given the AC voltage V5=1.0Vrms, predict the primary voltage and secondary voltage in the circuit shown in Figure 31-2. Record your result in Table 31-1.
1. Given the AC voltage V5=1.0Vrms, predict the power dissipation in resistor R_L2 in circuit shown in Figure 31-2. Record your result in Table 31-1.

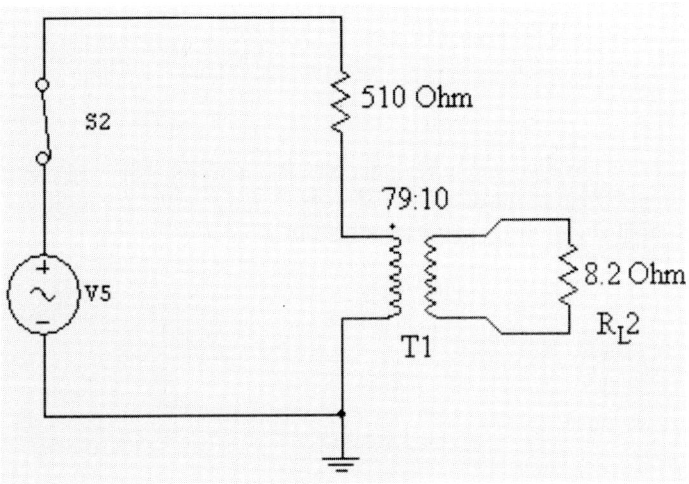

Figure 31-2

Lab Procedure

Before proceeding further, make sure that your eTCB is working correctly. Using eTCB Diagnostic VI, perform a self-diagnostic of eTCB circuitry.

Turn the switch S1 and S2 OFF in preparation for the lab.

Setup

Step #1 Using the eTCB Configuration VI, (AC module), configure the circuit in Figure 31-3

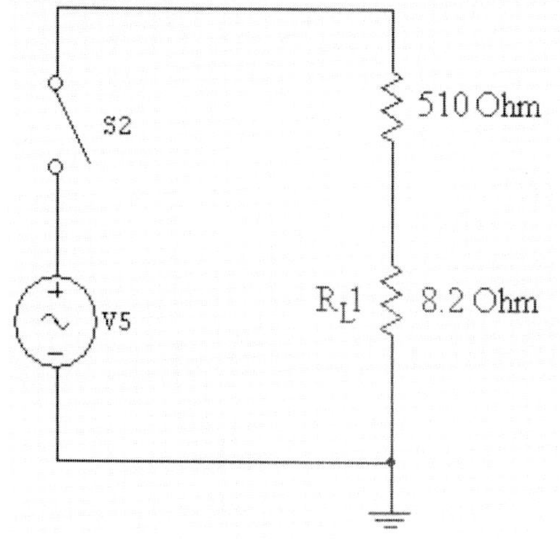

Figure 31-3

Step #2 Using a Function Generator VI, set voltage source V5 to ***2.0V*** peak amplitude at ***3.0k Hz***. Keep in mind that this setting is just a good indication of what the voltage of V5 is set to. It may not necessarily be the actual voltage of V5. Amplitude of signals from function generators often need to be adjusted (tuned) in order to set their output voltage precisely. Keep the Function Generator VI ***open***; if you close it, its output will reset to 0V.

Step #3 Physically, turn the switch S2 on the eTCB board ON. This will complete your circuit. Using the Digital Multimeter VI, measure voltage at **TP40** with respect to ground. Adjust the amplitude of the output of the function generator to obtain a reading of ***1.0Vrms.***

Discovery

Note: We are now going to perform voltage measurement to determine the power dissipated in the load resistor, R_L1. This will be done with a circuit which does not have an impedance matching transformer.

Step #4 Using the Digital Multimeter VI, measure the voltage across the R_L1 resistor. Record this reading in Table 31-1 (column – Measured).

Setup

Step #5 Using the eTCB Configuration VI, (AC module) configure the circuit in Figure 31-4

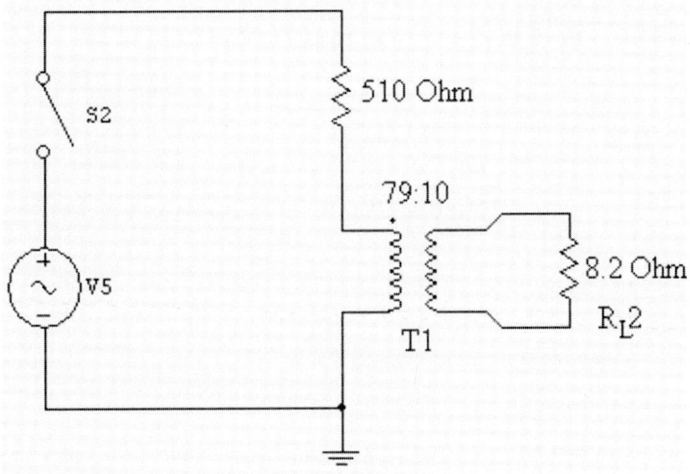

Figure 31-4

Step #6 Using a Function Generator VI, set voltage source V5 to ***2.0V*** peak amplitude at ***3.0k Hz***. Keep in mind that this setting is just a good indication of what the voltage of V5 is set to. It may not necessarily be the actual voltage of V5. Amplitude of signals from function generators often need to be adjusted (tuned) in order to set their output voltage precisely. Keep the Function Generator VI ***open***; if you close it, its output will reset to 0V.

Step #7 Physically, turn the switch S2 on the eTCB board ON. This will complete your circuit. Using the Digital Multimeter VI, measure voltage at **TP40** with respect to ground. Adjust the amplitude of the output of the function generator to obtain a reading of ***1.0Vrms.***

Discovery

Note: We are now going to perform voltage measurement to determine the power dissipated in the load resistor, R_L2. This will be done with a circuit which has an impedance matching transformer.

Step #8 Using the Digital Multimeter VI, measure the primary transformer voltage. Record this reading in Table 31-1 (column – Measured).

Step #9 Using the Digital Multimeter VI, measure the secondary transformer voltage, which is also the voltage across the ***R_L2*** resistor. Record this reading in Table 31-1 (column – Measured).

This concludes the procedure for this experiment.

Table 31-1

Circuit without Transformer		
	Predicted	***Measured***
Power dissipated in R_L1		*
Circuit with Transformer		
	Predicted	***Measured***
Primary Voltage		
Secondary Voltage		
Power dissipated in R_L2		*

	* Calculated power using measured voltage values.

Questions

1. Perform the error calculation for all entries in Table 31-1. Comment on whether the errors are acceptable.

1. Using measured primary and secondary voltage data from Table 31-1, calculate the turn ratio of the transformer in the circuit in Figure 31-5.

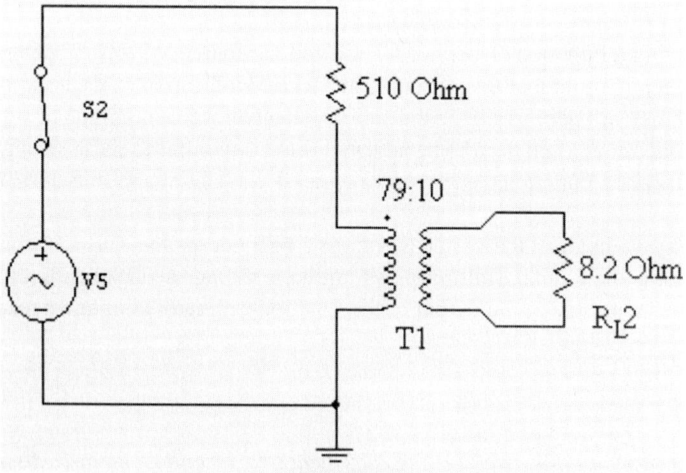

Figure 31-5

1. Using predicted and measured, primary and secondary voltage data from Table 31-1, calculate the equivalent primary resistance of the transformer circuit in Figure 31-5.

1. Using predicted and measured data from Table 31-1, calculate the primary and secondary power of the transformer circuit in Figure 31-5.

Bonus:

1. The given audio amplifier is known to be configured to drive an 8 _ speaker. The internal resistance of a speaker which is to be powered by that amplifier is 4 _. Draw the circuit in which an audio transformer is used to achieve a maximum power transfer from the amplifier to the 4 _ speaker. Make sure to determine the turn ration of the used transformer.

32

Coupling and Decoupling Capacitor Circuits

Problem Statement

Consider a scenario where a transducer converts a temperature into a DC voltage. The system is installed in the environment where a lot of high frequency signals are induced in the wires carrying DC voltage information from the transducer to the device where the DC voltage is being converted into a temperature reading. Your challenge is to build a circuit that would get rid of the unwanted AC signal which was coupled onto the DC voltage signal from the transducer. Understanding the behavior of the decoupling capacitor circuits will help you in solving this challenge.

Concept

A sinusoidal waveform may have a non-zero reference voltage. This voltage is referred to as a DC offset voltage. Figure 32-1 shows a sinusoidal wave form with a 5V DC offset. Signals like this can be created by **coupling** a sinusoidal waveform with 0V DC offset onto a 5V DC potential.

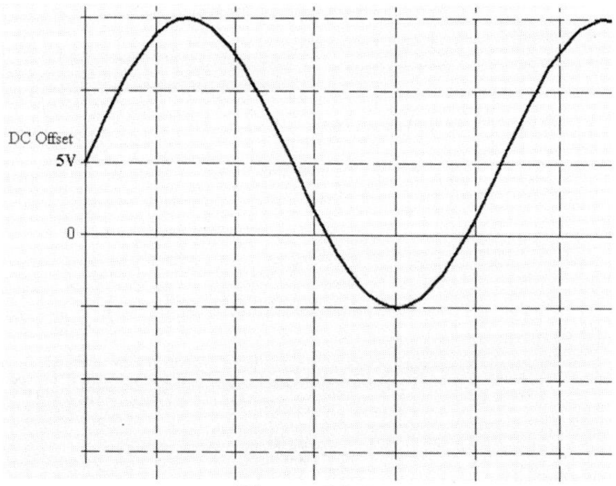

Figure 32-1

The coupling of the AC signal onto DC signal is achieved with use of coupling capacitor which, while conducting AC signal, isolates one DC reference voltage from another. Figure 32-2 illustrates the process of coupling an AC signal onto a DC potential.

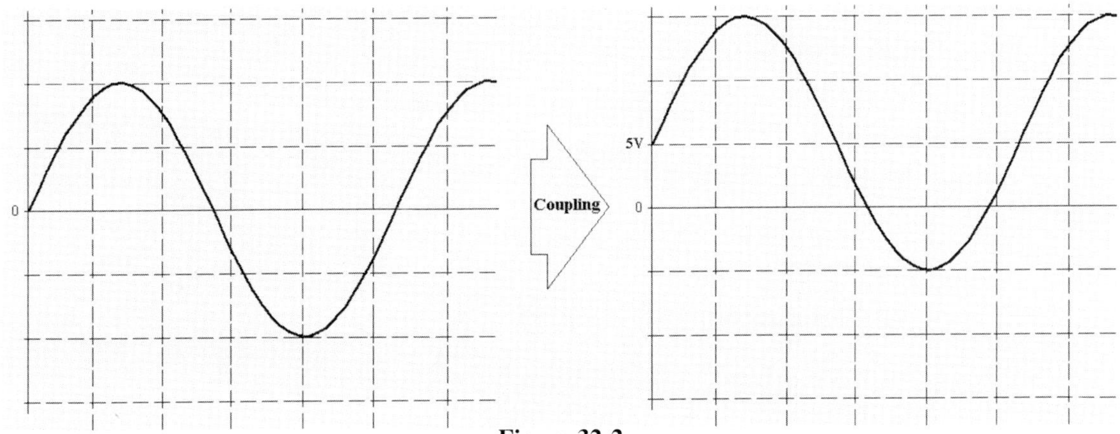

Figure 32-2

A **decoupling** capacitor is designed to create a low impedence path for AC signals without changing the DC voltages across the decoupling capacitor. The decoupling of an AC signal from a DC potential is achieved with the use of decoupling capacitor. Figure 32-3 illustrates the process of decoupling AC signal from a DC potential.

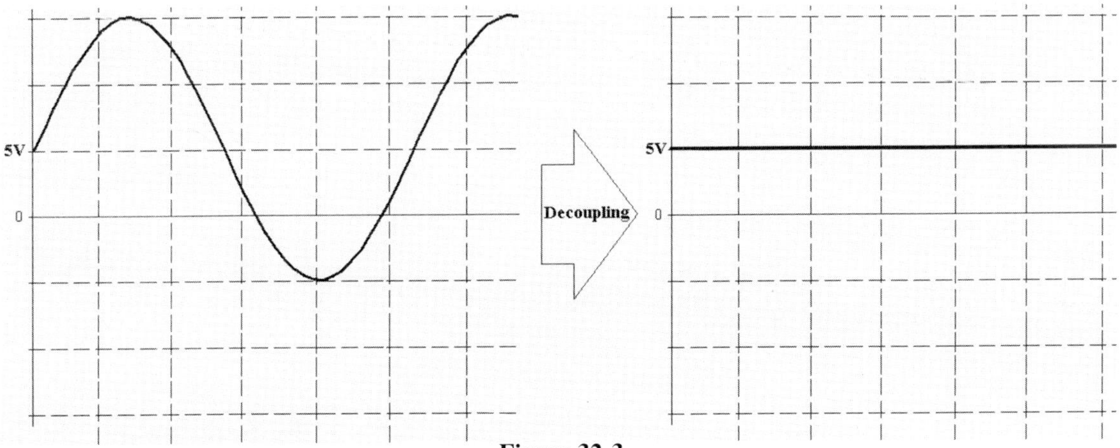

Figure 32-3

Learning Objectives

Upon completion of this lab, the student will be able to:

1. Measure AC signals before and after being coupled onto a DC potential.
1. Measure AC signals before and after being decoupled from a DC potential.

Pre-Lab Procedure

1. Predict the DC voltage at Point A in the circuit shown in Figure 32-4, V5=0V, V6=15V. Record your results in Table 32-1.

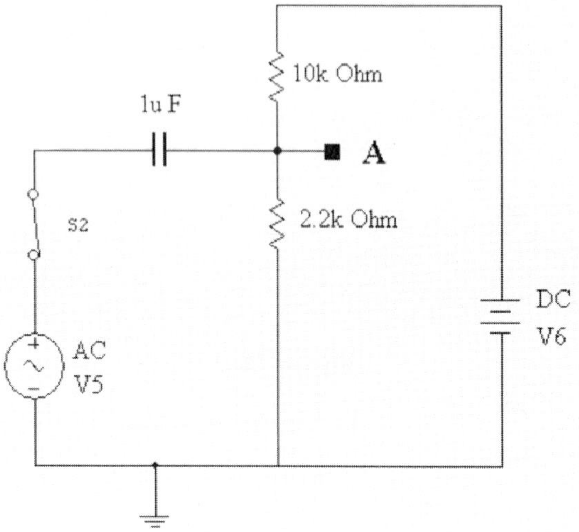

Figure 32-4

1. Redraw the circuits shown in Figure 32-4 for AC signals with very high frequencies. ($X_C=1/2\pi f$)

1. Redraw the circuits shown in Figure 32-5 for AC signals with very high frequencies. ($X_C=1/2\pi f$)

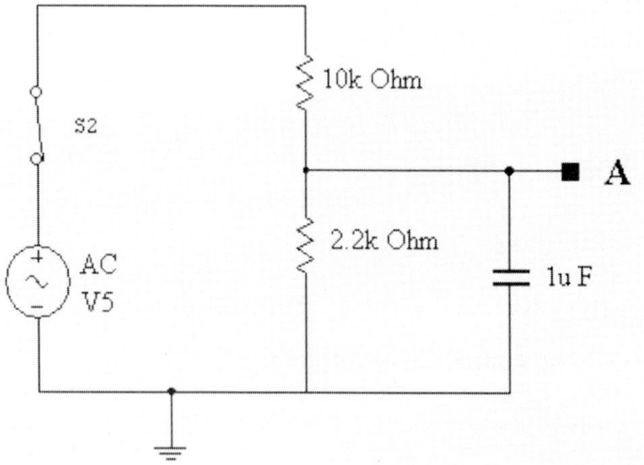

Figure 32-5

Lab Procedure

Before proceeding further, make sure that your eTCB is working correctly. Using eTCB Diagnostic VI, perform a self-diagnostic of eTCB circuitry.

Turn the switch S1 and S2 OFF in preparation for the lab.

Setup

Step #1 Using eTCB Configuration VI (AC module), reset all relays and then configure the circuit shown in Figure 32-6.

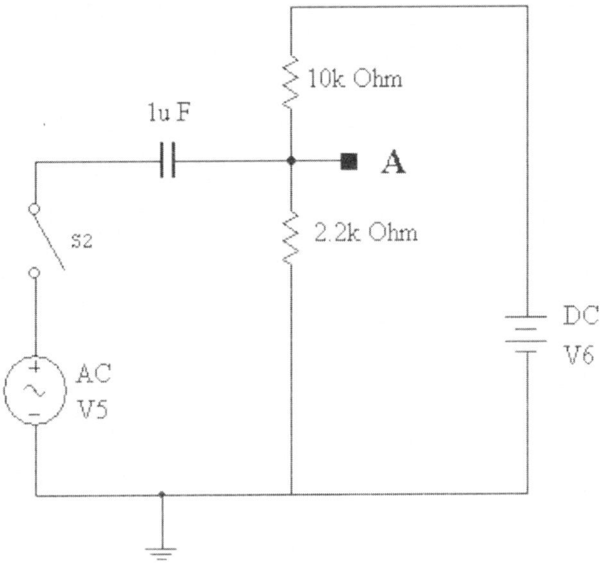

Figure 32-6

Step #2 Using a Function Generator VI, set the voltage source V5 to ***1.75V peak*** amplitude at ***5k Hz***. Keep in mind that this setting is just a good indication of what the voltage of V5 is set to. It may not necessarily be the actual voltage of V5. Amplitude of signals from function generators often need to be adjusted (tuned) in order to set their output voltage precisely. Keep the Function Generator VI ***open***; if you close it, its output will reset to 0V.

Step #3 Physically, turn the switch S2 on the eTCB board ON. This will complete your circuit. Use Oscilloscope VI, to measure voltage at TP40 with respect to ground. Display this signal on CHANNEL A. Use the measurement option on the oscilloscope to measure signal parameters on CHANNEL A. Continue to perform this measurement while adjusting the signal amplitude with function generator. Adjust V5 to ***3.5Vp-p***

Discovery

Note: We are now going to observe the effects of ***coupling*** *AC signal onto a DC potential.*

Step #4 Using an Oscilloscope VI, continue to measure the output of the function generator making sure that its output voltage remains at 3.5Vp-p. You may need to adjust it when it drifts too far away from 3.5Vp-p. Use CHANNEL B of the oscilloscope to measure the voltage at **Point A** (Figure 32-6). Record this reading in Table 32-1.

Setup

Step #5 Using eTCB Configuration VI (AC module), reset all relays and then configure the circuit shown in Figure 32-7.

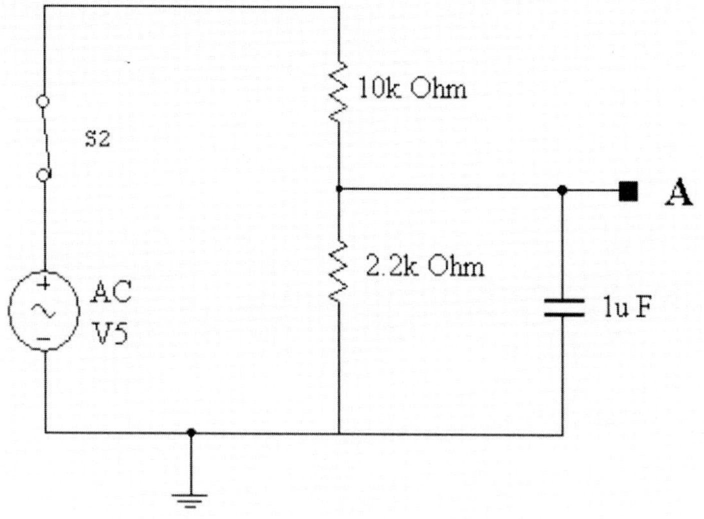

Figure 32-7

Discovery

Note: We are now going to observe the effects of ***decoupling*** *AC signal onto a DC potential.*

Step #6 Using an Oscilloscope VI, continue to measure the output of the function generator making sure that its output voltage remains at 3.5Vp-p. You may need to adjust it when it drifts too far away from 3.5Vp-p. Use CHANNEL B of the oscilloscope to measure the voltage at **Point A** (Figure 32-7). Record this reading in Table 32-1.

This concludes the procedure for this experiment.

Table 32-1

Predicted DC voltage at Point A	*Measured DC voltage at Point A*
Coupling Capacitor Circuit	
	Measured Voltage at ***Point A***
Peak-to-Peak Voltage	
DC offset Voltage	
RMS Voltage	
Signal Frequency	
Decoupling Capacitor Circuit	
	Measured Voltage at ***Point A***
Peak-to-Peak Voltage	
DC offset Voltage	
RMS Voltage	
Signal Frequency	

Questions

1. Has the frequency of the AC signal changed as the signal was coupled?

1. Explain why, when choosing the value of a coupling capacitor, is it important to consider the frequency of the AC signal to be coupled?

1. Explain why, when choosing the value of a decoupling capacitor, is it important to consider the frequency of the AC signal to be decoupled?

Bonus:

1. Why was the AC signal coupled slightly less then the AC signal generated by the function generator.

Appendix

Self-Diagnostic Codes

eTCB system has a built-in self-diagnostic feature. This feature is design to acquire a DC voltage measurement for various scenarios of input conditions and circuit configurations. For DC module DC voltage measurement is acquired between TP25 and TP20. By automatically controlling the status of various relays different circuit are configured. Given known circuit configurations and input Voltages, measurements of voltage between TP25 and TP20 are acquired and compared to the acceptable ones. In the event of a mismatch between the acquired voltage reading and acceptable one, an indication is given informing the user of the diagnostic step and the voltage reading at the time of the mismatch. This information can be used by the user to further determine the source of the mismatch.

Steps 1 to 20 are examining the DC portion of the eTCB module.

For steps (1 to 20) of the diagnostic procedure, DC voltage is measured always between TP25 and TP20. Variable power supplies are set at +9VDC and –9VDC.

The following illustrates the configuration for Step #1.

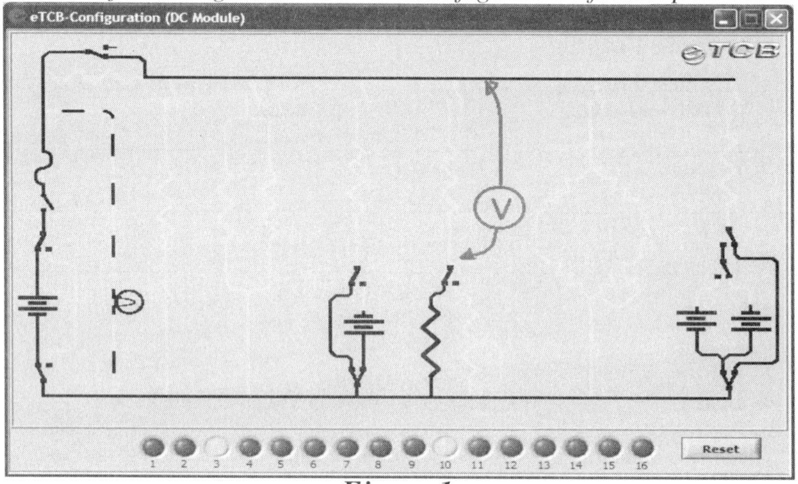

Figure 1

Steps	1	2	3	4	5	6	7	8	9	10	11	12	13	14	15	16
1			1							1						
2			1						1							
3	1								1	1						
4		1				1	1	1		1						
5		1				1	1	1	1	1						
6	1	1				1	1	1	1	1						
7	1	1				1	1		1	1						
8				1					1	1						
9	1			1					1	1						
10			1	1					1	1						
11					1				1	1						
12			1		1				1	1						
13									1	1	1					
14									1	1	1	1				
15									1	1			1			
16		1							1	1			1	1		1
17	1	1							1	1			1	1		1
18		1								1			1	1	1	1
19		1							1	1			1	1	1	1
20																

Steps 21 to 32 are examining the AC portion of the eTCB module.

This feature is design to acquire a AC voltage measurement for various scenarios of input conditions and circuit configurations. For AC module AC voltage measurement is acquired between GND and TP49. By automatically controlling the status of various relays different circuit are configured. Given known circuit configurations and input Voltages, measurements of AC voltage between GND and TP49 are acquired and compared to the acceptable ones. In the event of a mismatch between the acquired voltage reading and acceptable one, an indication is given informing the user of the diagnostic step and the voltage reading at the time of the mismatch. This information can be used by the user to further determine the source of the mismatch.

For steps (21 to 32) of the diagnostic procedure, RMS voltage is measured always between GND and TP49.

Vac is set at 1.45Vrms for all steps of the diagnostic.

The following illustrates the configuration for Step #21.

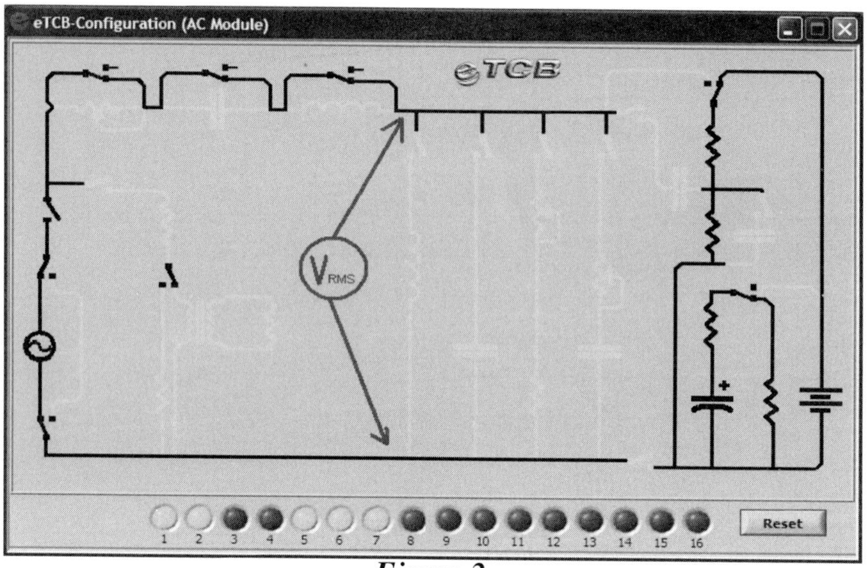

Figure 2

Steps	1	2	3	4	5	6	7	8	9	10	11	12	13	14	15	16	Freq [Hz]
21	1	1			1	1	1										1000
22	1	1				1	1			1							1000
23	1	1				1	1		1								1000
24	1	1				1	1		1								300
25	1	1				1	1	1									5000
26	1	1				1	1	1									3000
27	1	1				1	1				1						3000
28	1	1				1	1				1						100
29	1	1			1		1			1							1000
30	1	1			1		1			1							200
31	1	1			1	1				1							200
32	1	1			1	1				1							3000

NOTES

Notes

NOTES

NOTES

NOTES

Notes

NOTES

Notes